Routledge Revivals

Reform in New York City

Originally published in 1991, *Reform in New York City* provides an interpretive synthesis of urban progressivism and provides a comprehensive historical look at progressivism in New York City. The book argues that urban reform still poses a major historiographical challenge to historians working today and that there is limited analysis of the social and political action that characterised turn of the century New York. The book addresses the conceptual approaches, interpretive differences, and thematic emphasis of the urban reform agenda.

Reform in New York City

A Study of Urban Progressivism

By Augustus Cerillo

Routledge
Taylor & Francis Group

First published in 1991
by Garland

This edition first published in 2018 by Routledge
2 Park Square, Milton Park, Abingdon, Oxon, OX14 4RN
and by Routledge
711 Third Avenue, New York, NY 10017

Routledge is an imprint of the Taylor & Francis Group, an informa business

Publisher's Note
The publisher has gone to great lengths to ensure the quality of this reprint but points out that some imperfections in the original copies may be apparent.

Disclaimer
The publisher has made every effort to trace copyright holders and welcomes correspondence from those they have been unable to contact.

A Library of Congress record exists under LCCN: 91010854

ISBN 13: 978-1-138-49135-9 (hbk)
ISBN 13: 978-1-351-03318-3 (ebk)
ISBN 13: 978-1-138-49136-6 (pbk)

Modern American History

New Studies and Outstanding Dissertations

Edited by
Robert E. Burke
and
Frank Freidel

A Garland Series

REFORM IN NEW YORK CITY

A Study of Urban Progressivism

Augustus Cerillo, Jr.

Garland Publishing, Inc.
New York & London
1991

Library of Congress Cataloging-in-Publication Data

Cerillo, Augustus
Reform in New York City : a study in urban progressivism /
Augustus Cerillo, Jr.
p. cm. — (Modern American history)
Thesis (Ph. D.)—Northwestern University, 1969.
Includes bibliographical references and index.
ISBN 0-8240-1893-1 (alk. paper)
1. New York (N.Y.)—Social conditions—Case studies. 2. Progressivism (United States politics)—History—Case studies. 3. Urban policy—New York (N.Y.)—History—Case studies. I. Title. II. Series.
HN80.N5C45 1991
303.48'4'097471—dc20 91-10854

Design by Marisel Tavarez

For Dee

Lauri and Mike

INTRODUCTION

The publication of my doctoral dissertation on progressivism in New York City some twenty years after it was submitted to the History Department of Northwestern University offers me an opportunity to note some trends in recent writings on early twentieth century urban reform.[1] When I began my research in the early 1960s, the views of George Mowry and Richard Hofstadter still largely dominated progressive scholarship. Mowry and Hofstadter had redirected the study of progressivism toward the city, and, in what proved to be the most controversial, criticized and perhaps least enduring aspect of their work, located the wellsprings of progressivism in a status revolt by the nation's Protestant older middle class of professionals and independent businessmen.[2] Even as some scholars were testing the so-called "Mowry-Hofstadter" thesis in local and state contexts and finding it wanting, other historians, challenging it on the grounds of faulty methodology, insufficient evidence, or too narrow a focus, pointed the way toward alternative views of progressive origins, motivations, goals and achievements.[3] These competing interpretations had significant implications for historians studying urban reform.

The decade of the 1960s was hardly under way when Robert Wiebe and Gabriel Kolko each published a book focusing on the connection between businessmen and national economic regulation during the progressive era. Although they reached strikingly different conclusions about the nature, power and goals of the business community, they both demonstrated the key role of businessmen in shaping federal legislation regulating business. Their findings suggested possibilities for a fresh examination of who or what social and economic groups exercised the most influence over municipal policymaking.[4]

The year after Kolko's book appeared, Samuel P. Hays published his often reprinted "The Politics of Reform in Municipal Government in the Progressive Era." In this essay he located the social sources of urban progressivism in a new upper class of banking and industrial leaders and advanced professionals who, he wrote, sought "to take formal political power from the previously dominant lower- and middle-class elements so that they might advance their own conceptions of desirable public policy."[5] At about the same time Wiebe, Kolko and Hays were focusing their studies of progressivism on middle- and upper-class elites, historian J. Joseph Huthmacher in two pioneering essays showed how urban foreign-born and working-class groups and their political leaders, including machine bosses, provided critical support for many progressive reforms.[6] Anticipating a later theme in progressive historiography Huthmacher concluded, "The coalition nature of the impulse behind the reform spirit of the Progressive Era was its greatest strength."[7]

By the mid-1960s then, the "progressive club" had opened to include, along with Mowry's and Hofstadter's old middle class, Wiebe's middle class "businessmen on the make,"[8] Hay's and Kolko's upper class elites and Huthmacher's immigrant and working class leaders. Such was the historiographical ferment surrounding progressive scholarship when I began my study of progressivism in New York City. I sought to discover what reform issues engaged New Yorkers in the first two decades of the new century and, in light of the debate over the origins and content of progressivism, to search for who and what groups played the leading roles in formulating and implementing Gotham's reform agenda.

At the time I began my research, Wiebe was already working on <u>The Search For Order</u>, which, after its publication in 1967, quickly became the most comprehensive and influential formulation of what others would label the "organizational" interpretation of progressivism. Appearing in print when I was well into the writing of "Reform in New York City," <u>The Search For Order</u> helped me sharpen the conceptual framework of my study. Wiebe viewed urban progressivism as part of a broader effort by a new middle class of professional and business groups to order and improve urban-industrial society through organizational and bureaucratic means.[9] This was consistent with what I discovered about much of the leadership in New York reform. Although I initially had no intention of doing a comparative study of Gilded Age and progressive era reform in New York City, my search for the roots of the movements for better utility services and regulation, health and housing laws and the renovation of government and politics, the issues I limited my study to, inevitably pushed me back into the nineteenth century. Influenced by Wiebe's stress on the discontinuity of reform on both sides of the century, I perceived a rather sharp break in the style, sources and content of reform in New York City around 1900.[10]

Since the late 1960s historians have continued the search for the origins, content and meaning of progressivism, and in the process have published scores of books and articles on the subject. There are a number of excellent essays that survey, interpret and critique this literature and I need not duplicate their content here.[11] Instead, for the remainder of this short introduction I will discuss a few historiographical trends that I believe have pertinence to the themes of this book as well as the study of urban progressivism.

During the 1970s and early 1980s, historians who continued to search for the political impulses of progressivism showed just how diverse were the sources of reform. In a number of essays and an important book, John Buenker fleshed out Huthmacher's earlier sketch of the role of the "urban foreign stock working class" in reform. He insisted immigrant workers, like the middle and upper classes, organized to protect and pursue their interests. Their political representatives supported governmental measures designed to regulate business, especially utilities, assure a fair tax rate, democratize the political system and provide for social welfare; they fought against certain "cultural reforms" such as Sunday closing and anti-gambling laws, and prohibition. Along with Huthmacher, Buenker also placed the label "urban liberalism" on this package of reforms and saw them as the precursors of the welfare state.[12] Other historians have attached the label "social reform" to such measures and the title "social reformer" to those leaders, such as Detroit's Mayor Hazen S. Pingree, who championed a reform program designed to meet the social, economic, and political needs of the city's working classes and to secure a more just and democratic society.[13]

Huthmacher's and Buenker's stress on the urban workers' support for reform was reinforced by the scholarship of a number of urban historians who mounted a full-scale re-evaluation of urban bosses and machines. Focusing on the functions of the political machine, and in some cases even de-emphasizing the traditional distinction between bossism and reform, these revisionists argued that the political boss at times supported, if he did not initiate, a variety of reform measures. Some of these historians even transformed old villains (bosses) into heroes (reformers) who championed the cause of the urban masses. Bosses, these urban scholars pointed out, gave welfare services to the poor, centralized decision making in a fragmented and disorderly city, supported development projects, extended democracy to the grassroots and fended off Yankee-Protestant middle-class reform assaults on immigrant morality. Of course, not all historians accepted the revisionists' overly "romantic" view of city bosses, and more recently some historians have challenged the historical accuracy and conceptual soundness of the functional view of the boss and political machine. Nevertheless, this scholarship forces historians of urban progressivism to take into account the part bosses played in the structure of urban rule, the reform process and more generally the politics of the emerging modern city.[14] In <u>Boss Cox's Cincinnati: Urban Politics in the Progressive Era</u>, one of the few studies of a single city that attempted to place both bossism and reform within the larger "process of urbanization," Zane Miller concluded that both bossism and reform "were interlocking parts in a new system of urban politics."[15]

Miller and his doctoral mentor, urban historian Richard Wade, added a spatial dimension to the study of urban politics, pointing out that the boss and his constituents resided in the urban core while the reformer and his followers mainly lived in the newer outlying areas of the expanding metropolis. Reform, in Wade's words, was "a movement of the periphery against the center."[16] In his study of Cincinnati, Miller elaborated on the implications of such a sociospatial interpretation of urban politics: "The pattern of politics suggests that residence rather than race, religion, or ethnicity provides the touchstone to the city's social and political experience."[17] Bruce Stave suggested a modification of the "periphery versus center" interpretation of bossism and reform. "Rather than a conflict between periphery and center," he wrote, "the dispute between urban bosses and reformers might be seen in terms of a contest between the center-as-residence (for the bosses and their immigrant following) and the center-as-place-of-business (for the reformers)."[18]

At about the same time historians were elevating city bosses and their working class constituents to the rank of reformers, David P. Thelen, in several provocative works, argued that in Wisconsin, and by implication elsewhere, the reform impulse stemmed not from a particular class victimized by social tensions and insecurities but from a broad coalition of groups who cooperated across class, ethnic and religious lines as citizens, taxpayers and consumers. Catalyzed by the depression of 1893, such groups of "new citizens" attacked corporate arrogance, unfair tax systems, undemocratic political practices and other social ills. "The basic riddle in progressivism," he wrote, "is not what drove groups apart, but what made them seek common cause."[19]

By the mid-1970s, historians Peter Filene, David Thelen and John Buenker confronted this mounting evidence that progressivism derived from more complex sources than had earlier been imagined by proposing a "coalition theory" to account for the diverse sources, ideas and issues that comprised progressivism.[20] This approach to the problem of diversity in fact reached back several years to a view articulated by Samuel P. Hays. In The Response to Industrialism, a widely read interpretation of the years 1885-1914, Hays largely ignored the category of "progressive" and instead described the progressive era as a time when all sorts of Americans organized "to cope with industrial change in all its ramifications."[21] Now this new generation of historians also denied there existed a single reform type or more generally a unified progressive movement. "The progressive era," wrote Filene in "An Obituary for 'The Progressive Movement,'" "seems to be characterized by shifting coalitions around different issues, with the specific nature of these coalitions varying on federal, state and local levels, from region to region, and from the first to the second decades of the century."[22]

Unlike Filene, who in the face of such diversity and complexity pessimistically asked historians to abandon the notion of a progressive movement, Buenker more optimistically saw in the coalition theory "the potential for reconciling most of the currently conflicting interpretations and of encompassing nearly all of the groups, values and programs that were plainly at work" in the progressive era. Such loose, shifting coalitions, Buenker argued, "are, by definition, temporary. They are not limited to any single social type. They require no fundamental agreement upon values or motives; factions with significantly different outlooks on life can usually find limited areas of accommodation and cooperation." Furthermore, Buenker stated, a coalition approach to reform "dismisses no group from participation because it fails to meet preconceived standards of class, ethnicity, or belief."[23] Much like those political scientists who advocated a "pluralist" view of the exercise of political power, historians now asserted that to understand progressivism one must examine each reform issue and the distinct coalition behind it.[24]

The current dominance of the coalition approach to analyzing progressivism notwithstanding, historians Thelen and

Richard L. McCormick have warned historians not to distort their understanding of progressivism by focusing too exclusively on reform leaders. As Thelen graphically wrote in his study of Wisconsin progressivism, "The voters generated the progressive programs and, familiar with the devices by which older politicians had betrayed them, they insisted that progressive leaders keep their tent flaps open so that the voters could observe their methods and, if need be, discipline them." "The early Wisconsin Progressives," he contended, "put more faith in developing new voluntary associations and the techniques of mass, popular, grass-roots pressure than they put in their leaders."[25] In his popular text on progressivism, co-authored with Arthur S. Link, McCormick asserted that "the dynamics of progressivism were crucially generated by ordinary people--by the sometimes frenzied mass supporters of progressive leaders, by rank-and-file voters willing to trust a reform candidate."[26] McCormick attributed the timing of such support for progressive reforms to a series of precipitating crises stemming from revelations across urban America of business corruption of politics and government.[27]

Not only did McCormick seek to bring ordinary urban dwellers back into the progressive story, he also challenged students of progressivism not to abandon the effort "to appreciate and understand Progressivism as a whole,"[28] or to search for those "features which were more or less common to different progressive reforms."[29] McCormick, as well as other historians, have insisted that in spite of the diversity of personnel and program, much of progressivism was characterized by a distinctive package of ideas, values and attitudes. Progressives in the main accepted industrialism and the need to ameliorate its worst consequences. They were optimistic about their ability to modify the social, economic and political environment to achieve individual well being and a more humane, just and orderly society. They relied on private voluntary efforts and most importantly coercive government power to intervene in people's personal lives and to improve the conditions of urban industrial life. They had great faith in the social sciences and sought to apply its methods to solving urban problems. And many reformers were energized by an Evangelical-inspired moral earnestness to purge society of "evil" and in some cases impose "a Protestant social order" on the community.[30]

Historians might disagree about how pervasive this set of beliefs was among the diverse groups of reformers, but few could doubt the impact of combinations of these ideas on at least portions of the urban progressive agenda. Even Buenker, who rejects the notion of "a common set of values" among the reformers, nevertheless concedes that they shared a "few very general convictions," one of the most significant being a belief "that government at all levels had a responsibility to promote human welfare." He previously had pointed to government intervention in the economy and the creation of a welfare state as important components of "urban liberalism." Of the progressive era as a whole, Buenker later wrote, "This widespread consensus that the state should actively promote social justice clearly marked the period off from earlier ones and established a precedent for future reformers."[31]

In an essay I contributed to The Age of Urban Reform edited by Michael H. Ebner and Eugene M. Tobin, I examined the interaction of ideology and reform in the thoughts and activities of several New York progressive leaders. Whatever their reform specialty, they all shared a common faith in administered progress and the bureaucratic management of urban problems.[32] More recently, Terrence J. McDonald, in an imaginative examination of the politics of fiscal policy in late nineteenth and early twentieth century San Francisco, showed how that city's progressive leaders broke sharply with the low tax, low expenditures and no public sector growth ideology of the nineteenth century and led pro-growth reform coalitions of businessmen, labor and neighborhood groups behind public sector solutions to their problems.[33]

Other historians have approached the issue of urban rule and public policy making by studying the sources, distribution and use of power in American cities. Books such as David Hammack's Power and Society: Greater New York at the Turn of the Century[34] and Carl V. Harris's Political Power in Birmingham, 1871-1921[35] demonstrate that cities in the industrial age were, in Raymond Mohl's words, "shaped by the continual political interaction of competing elites, pluralistic interest groups, and urban technicians."[36] In his outstanding volume, Hammack examined four areas of conflict in turn-of-the-century New York City: mayoral politics, the creation of Greater New York, the construction of the city's first subway, and the centralization of the public school system. "The evidence indicates," he concluded,

> that power was concentrated not in one or two but in several distinct economic, social, and political elites. In their efforts to influence decisions these elites engaged in a shifting complex of alliances, bargained with one another, and sometimes made important concessions to secure the support of other elites and of wider publics. The city was neither ruled by a single power elite, nor by a genteel, patrician, or merely rich upper class. Nor was it ruled by an all-powerful political machine. Middle- and lower-income New Yorkers influenced decisions directly, through a variety of increasingly effective organizations, including several political machines and a variety of special-interest groups, and indirectly, through their votes, petitions, demonstrations, strikes, and boycotts.[37]

Studies such as Harris' and Hammack's have been criticized for being too narrow in scope or for ignoring the impact on municipal decisions of institutional structures, procedures and customs.[38] Whatever their shortcomings, they have provided additional evidence for a view of reform that stresses diversity and shifting coalitions as the key to understanding urban progressivism.

Recently Jon Teaford has been a leader among those historians calling for more scholarly attention to the "neglected participants in urban government."[39] In two immensely helpful essays assessing the literature on urban rule, he convincingly argued the case for more studies on the precise role in decision making of public officials such as mayors, comptrollers, city engineers, attorneys and an assortment of professional bureaucrats and administrators; of outside consultants, including landscape architects and planners; and of interest groups such as labor unions and fire insurance underwriters. He also urged more research on the expansion and development of municipal functions, services and public works. Such developments, he noted, opened within the city government new bureaucratic and technical occupational niches and places of leadership, and gave rise to a multiplicity of related outside pressure groups. In addition, Teaford cited the need for more information about the impact of municipal expenditures, taxation, and debt on governmental structures and public policy decisions.[40] "Historians have written blithely about waste, incompetence, and inefficiency in city government," he wrote, "without ever testing these generalizations against the municipal ledgers."[41] More generally, he concluded, the story of urban rule and by implication urban reform is "a story of greater breadth and complexity than has yet been recorded."[42]

In his splendid book, The Unheralded Triumph: City Government in America, 1870-1900, Teaford provided ample evidence of the role of city bureaucrats, technicians and trained experts as well as an assortment of businessmen, neighborhood groups and politicians--"the various forces pushing and pulling the municipal corporations"--in shaping urban government and policy.[43] "Patricians, business leaders, engineers, librarians, partisan ward bosses, grocers, and saloonkeepers," he wrote, "all had their niche in urban government, and any major project or achievement involved a meeting of two or more of these diverse and sometimes hostile elements. . . . No single group or person pulled the strings of city government."[44] Teaford also included a discussion of the impact of fiscal policies, ideologies and constraints on Gilded Age urban politics and infrastructural development, a central theme of Terrence J. McDonald's recent book on San Francisco.[45]

Teaford's book also challenged the prevailing view that late nineteenth century city governments failed to cope successfully with the problems of rapid physical and industrial growth and increased social and economic complexity. He described the use of trained experts, the reduction of ethnocultural tensions, the provision of urban services and the political accommodation of diverse interest groups as "grand achievements" of late nineteenth century governments. This "unheralded triumph" of Gilded Age municipal rule, Teaford stated, bequeathed to the progressive generation a positive legacy of "monuments of public enterprise" and "the framework of urban rule."[46] Teaford's analysis forces historians to reconsider the originality of later urban progressive programs and methods of reform, and thus to re-examine the precise connection between urban reform on both sides of the century.

Whatever their conceptual approaches, interpretive differences and thematic emphases, historians generally agree on the broad categories of issues that constituted the urban reform agenda, and have published an abundance of scholarly works that has greatly enriched our understanding of which individuals and groups sought what types of reforms for what reasons with what results. Historians have written about the structural renovation as well as democratization of government and politics; the promotion and regulation of urban services and business enterprises; social welfare, housing, health and labor measures; educational reform; zoning and city planning; the provision of recreational and other amenities; and a variety of moral reforms.[47]

Remarkably, no one has yet constructed out of these "historical bricks" a book-length interpretive synthesis of urban progressivism.[48] Nor for that matter has a historian yet published a comprehensive history of progressivism in the nation's premier city. We still have only more limited forays into the vastness of the social and political action that characterized turn-of-the-century New York. Thus the age of urban reform still poses a major historiographical challenge to today's historians.

Augustus Cerillo, Jr.
Irvine, California
1990

NOTES

1. Augustus Cerillo, Jr., "Reform in New York City: A Study of Urban Progressivism," Ph.D. diss., Northwestern University, 1969.

2. George E. Mowry, The California Progressives (Berkeley, CA, 1951) and The Era of Theodore Roosevelt, 1900-1912 (New York, 1958); Richard Hofstadter, The Age of Reform: From Bryan to F.D.R (New York, 1955), pp. 131-271.

3. Richard B. Sherman, "The Status Revolution and Massachusetts Progressive Leadership," Political Science Quarterly 78 (1963), 59-65; William T. Kerr, Jr., "The Progressives of Washington, 1910-1912," Pacific Northwest Quarterly 55 (1964), 16-27; E. Daniel Potts, "The Progressive Profile in Iowa," Mid-America 47 (1965), 257-268; David P. Thelen, "Social Tensions and the Origins of Progressivism," Journal of American History 56 (1969), 323-341; Samuel P. Hays, "The Politics of Reform in Municipal Government in the Progressive Era," Pacific Northwest Quarterly 55 (October, 1964), 157-169; For a recent summary of the criticisms of Hofstadter see Alan Brinkley, "Richard Hofstadter's The Age of Reform: A Reconsideration," Reviews in American History 13 (September, 1985), 462-480, and for a critique of the critics see Jerome M. Clubb and Howard W. Allen, "Collective Biography and the Progressive Movement: The 'Status Revolution' Revisited," Social Science History, 1 (Summer, 1977), 518-534.

4. Robert H. Wiebe, Businessmen and Reform: A Study of the Progressive Movement (Cambridge, Mass., 1962); Gabriel Kolko, The Triumph of Conservatism (New York, 1963). In The Corporate Ideal in the Liberal State, 1900-1918 (Boston, 1968), James Weinstein examined the movements for workman's compensation and the city commission and council manager forms of local government. He found business leaders dominated both reform efforts and generalized that by the end

of the progressive years "business organizations were able rapidly to press forward their political domination of American municipalities." (116)

5. Quoted on p. 162; Also see Hays' "The Changing Political Structure of the City in Industrial America," Journal of Urban History 1 (November, 1974), 6-38, for his analysis of urban reform within the context of urbanization and its accompanying sociospatial and economic changes. He again located the sources of reform in the "central-city businessman, the advanced professional, and upper social classes," (p. 22) and interpreted their reform activities as the "triumph of centralizing over decentralizing forces" (p. 24) in urban life. For Hays' pioneering contributions to the social analysis of political history, see the essays collected in his American Political History as Social Analysis (Knoxville, TN, 1980). For a summary and critique of this approach to political history, see Richard L. McCormick, "The Social Analysis of American Political History--After Twenty Years" in The Party Period and Public Policy: American Politics from the Age of Jackson to the Progressive Era (New York, 1986), pp. 89-140.

6. "Urban Liberalism and the Age of Reform," The Mississippi Valley Historical Review, 49 (September, 1962), 231-241; "Charles Evans Hughes and Charles Francis Murphy: The Metamorphosis of Progressivism," New York History, 46 (January, 1965), 25-40.

7. "Hughes and Murphy," p. 36.

8. Wiebe, Businessmen and Reform, pp. 209, 210.

9. (New York); Louis Galambos, "The Emerging Organizational Synthesis in Modern American History," Business History Review 44 (Autumn, 1970), 279-290. For an insightful critique of the organizational synthesis see Wayne K. Hobson, "Professionals, Progressives and Bureaucratization: A Reassessment," The Historian 39 (August, 1977), 639-658. Thomas Bender has critically evaluated Wiebe's concept of the breakdown of community as America passed from the nineteenth into

the twentieth century in Community and Social Change in America (Baltimore, 1978), pp. 45-58.

10. The Search for Order, pp. 164-176.

11. Peter G. Filene, "An Obituary for 'The Progressive Movement,'" American Quarterly 22 (1970), 20-34; William G. Anderson, "Progressivism: An Historiographical Essay," The History Teacher 6 (May, 1973), 427-452; Robert H. Wiebe, "The Progressive Years, 1900-1917," in The Reinterpretation of American History and Culture, eds. William H. Cartwright and Richard L. Watson, Jr. (Washington, D.C., 1973), pp. 425-442; David M. Kennedy, "Overview: The Progressive Era," The Historian 37 (May, 1975), 453-468; Daniel T. Rodgers, "In Search of Progressivism," Reviews in American History 10 (December, 1982), 113-132; Blaine A. Brownell, "Interpretations of Twentieth-Century Urban Progressive Reform," in Reform and Reformers in the Progressive Era, eds. David R. Colburn and George E. Pozzetta (Westport, Conn., 1983), pp. 3-23; For a useful survey of the literature on urban history see Michael H. Ebner, "Urban History: Retrospect and Prospect," The Journal of American History 68 (June, 1981), 69-84.

12. "The Progressive Era: A Search for a Synthesis," Mid-America 51 (July, 1969), 175-193; Urban Liberalism and Progressive Reform (New York, 1973); John D. Buenker, John C. Burnham and Robert M. Crunden, Progressivism (Cambridge, Mass., 1977), pp. 31-69. Roger E. Wyman in "Middle-Class Voters and Progressive Reform: The Conflict of Class and Culture," The American Political Science Review 68 (June, 1974), 488-504, shows that in Wisconsin ethnocultural and class factors were primary in determining voter choice in elections; he also asserts that progressivism was not a middle-class or upper middle-class movement in that state.

13. Melvin G. Holli made the widely adopted distinction between structural reform and social reform in Reform in Detroit: Hazen S. Pingree and Urban Politics (New York, 1969), and in "Urban Reform in the Progressive Era," in The Progressive Era, ed. Lewis L. Gould (Syracuse, 1974), pp. 133-

151. For a recent textbook that uses these categories see Raymond A. Mohl, The New City: Urban America in the Industrial Age, 1860-1920 (Arlington Heights, IL, 1985), pp. 115-128.

14. Alexander B. Callow, Jr., The Tweed Ring (New York, 1966); Lyle Dorsett, The Pendergast Machine (New York, 1968); Zane Miller, Boss Cox's Cincinnati: Urban Politics in the Progressive Era (New York, 1968); John M. Allswang, Bosses, Machines, and Urban Voters, Rev. Ed. (Baltimore, 1986); Nancy Weiss, Charles Francis Murphy, 1858-1924: Respectability and Responsibility in Tammany Politics (Northampton, Mass., 1968). M. Craig Brown and Charles N. Halaby, "Machine Politics in America, 1870-1945," Journal of Interdisciplinary History 17 (Winter, 1987), 587-612, questions accepted generalizations about the power and longevity of political machines while Gerald Kurland, Seth Low: The Reformer in an Urban and Industrial Age (New York, 1971), pp. 193-214, still prefers to see bosses as scoundrels. For helpful discussions of the issues related to the theme of bosses and machines see Lyle W. Dorsett, "The City Boss and the Reformer: A Reappraisal," Northwest Quarterly 63 (October, 1972), 150-154; David P. Thelen, "Urban Politics: Beyond Bosses and Reformers," Reviews in American History 7 (September, 1979), 406-412; Jon C. Teaford, "Finis for Tweed and Steffens: Rewriting the History of Urban Rule," Reviews in American History 11 (December, 1982), 133-149 and "New Life for an Old Subject: Investigating the Structure of Urban Rule," in American Urbanism: A Historiographical Review, eds. Howard Gillette, Jr., and Zane Miller (New York, 1987); Bruce M. Stave, John M. Allswang, Terrence J. McDonald and Jon C. Teaford, "A Reassessment of the Urban Political Boss: An Exchange of Views," The History Teacher 21 (May, 1988), 293-312. Three helpful anthologies on bosses and reformers are Blaine A. Brownell and Warren E. Stickle, eds., Bosses and Reformers (Boston, 1973); Alexander B. Callow, Jr., ed., The City Boss in America (New York, 1976); and Bruce M. Stave and Sondra

Astor Stave, eds., Urban Bosses, Machines, and Progressive Reformers, Second ed. (Malabar, FL, 1984).

15. P. 241.

16. Richard C. Wade, "Urbanization," in The Comparative Approach to American History, ed. C. Vann Woodward (New York, 1968), p. 197.

17. Boss Cox's Cincinnati, p. 241.

18. "Urban Bosses and Reform," in The Urban Experience: Themes in American History, eds. Raymond A. Mohl and James F. Richardson (Belmont, CA, 1973), p. 188.

19. "Social Tensions," p. 341; The New Citizenship: Origins of Progressivism in Wisconsin, 1885-1900 (Columbia, MO, 1972).

20. Buenker, Urban Liberalism, pp. 217-218.

21. (Chicago, 1957), quote is on p. 188; Buenker, Urban Liberalism, pp. 218-220.

22. P. 33.

23. Progressivism, quotes are on pp. 31 and 35.

24. Arthur S. Link and Richard L. McCormick, Progressivism (Arlington Heights, IL, 1983), p. 10.

25. The New Citizenship, p. 3.

26. Progressivism, p. 8.

27. Ibid., pp. 8-9; McCormick, "The Discovery That Business Corrupts Politics: A Reappraisal of the Origins of Progressivism," American Historical Review 86 (April, 1981), 247-274, also reprinted in The Party Period and Public Policy, pp. 311-356.

28. "Progressivism: A Contemporary Reassessment," in Party Period and Public Policy, p. 268.

29. Link and McCormick, Progressivism, p. 10.

30. This summary of progressive ideas is drawn from the following: Wiebe, The Search for Order, pp. 145-163, 169-176; Otis A. Pease, "Urban Reformers in the Progressive Era: A Reassessment," Pacific Northwest Quarterly 61 (April, 1971), 49-58; McCormick, "Contemporary Reassessment," pp. 269-272; Link and McCormick, Progressivism, pp. 21-25, 67-104; Clyde

Griffen, "The Progressive Ethos," in The Development of American Culture, eds. Stanley Coben and Norman Ratner (Englewood Cliffs, NY, 1970), pp. 120-149; Henry May, Ideas, Faiths, and Feelings (New York, 1983), pp. 163-186; Don S. Kirschner, "The Ambiguous Legacy: Social Justice and Social Control in the Progressive Era," Historical Reflections 2 (Summer, 1975), 69-88; Michael H. Frisch, "Urban Theorists, Urban Reform, and American Political Culture in the Progressive Period," Political Science Quarterly 97 (Summer, 1982), 295-315; R. Laurence Moore, "Directions of Thought in Progressive America," in The Progressive Era, ed. Gould, pp. 35-54; John Whiteclay Chambers II, The Tyranny of Change: America in the Progressive Era, 1900-1917 (New York, 1980), pp. 107-113; John C. Burnham, "Essay," in Progressivism, eds. Buenker, et al, pp. 3-29; and Rodgers, "In Search of Progressivism," pp. 122-127.

31. Progressivism, pp. 33, 49, 63; John C. Burnham, however, claims that progressives were at best ambivalent about the use of governmental power to achieve reform. See Burnham, "Essay," in Progressivism, pp. 14-16.

32. "The Impact of Reform Ideology: Early Twentieth-Century Municipal Government in New York City" (Port Washington, NY, 1977), pp. 68-85.

33. The Parameters of Urban Fiscal Policy: Socioeconomic Change and Political Culture in San Francisco, 1860-1906 (Berkeley, CA, 1986).

34. (New York, 1987). Also see David C. Hammack, "Problems in the Historical Study of Power in the Cities and Towns of the United States, 1800-1960," The American Historical Review, 83 (April, 1978), 323-349.

35. (Knoxville, 1977).

36. "New Perspectives on American Urban History," in The Making of Urban America, ed. Raymond A. Mohl (Wilmington, 1988), p. 299.

37. Power and Society, p. 304.

38. Teaford, "New Life for an Old Subject," pp. 98-100.

39. "Rewriting the History of Urban Rule," p. 137.

40. "Rewriting the History of Urban Rule," pp. 143-144; "New Life for an Old Subject," pp. 94-97.

41. "Rewriting the History of Urban Rule," p. 143.

42. Ibid., p. 147.

43. Quote on p. 8.

44. Ibid., pp. 308-309.

45. Ibid., pp. 284-306; McDonald, Urban Fiscal Policy.

46. The Unheralded Triumph, quotes on pp. 313, 312.

47. Ernest S. Griffith, A History of American City Government: The Progressive Years and Their Aftermath, 1900-1920 (Washington, D.C., 1983); Martin J. Schiesl, The Politics of Efficiency: Municipal Administration and Reform in America: 1880-1920 (Berkeley, 1977); Bradley R. Rice, Progressive Cities: The Commission Government Movement in America, 1901-1920 (Austin, 1977); Kenneth Fox, Better City Government: Innovation in American Urban Politics, 1850-1937 (Philadelphia, 1977); Dewey W. Grantham, Southern Progressivism: The Reconciliation of Progress and Transition (Knoxville, 1983), pp. 276-290; Bruce M. Stave, ed., Socialism and the Cities (Port Washington, NY, 1975); Michael H. Ebner and Eugene M. Tobin, eds., The Age of Urban Reform (Port Washington, NY, 1977); Charles W. Cheape, Moving the Masses: Urban Public Transit in New York, Boston, and Philadelphia, 1880-1912, (Cambridge, 1980); Anthony Jackson, A Place Called Home: A History of Low-Cost Housing in Manhattan (Cambridge, 1976); John Duffy, A History of Public Health in New York City, 1866-1966 (New York, 1974); Stuart Galishoff, Safeguarding the Public Health: Newark, 1895-1918 (Westport, Conn., 1975); Martin V. Melosi, Garbage in the Cities: Refuse, Reform, and the Environment, 1880-1980 (College Station, TX, 1981); Walter I. Trattner, From Poor Law to Welfare State: A History of Social Welfare in America (New York, 1979); Allen F. Davis, Spearheads for Reform: The Social Settlements and the Progressive Movement, 1890-1914 (New York, 1967); Thomas Lee Philpott, The Slum and the

Ghetto: Neighborhood Deterioration and Middle-Class Reform, Chicago, 1880-1930 (New York, 1978); Melvyn Dubofsky, Industrialism and the American Worker, 1865-1920, Second Ed. (Arlington Heights, IL, 1985); Roy Rosenzweig, Eight Hours for What We Will: Workers and Leisure in an Industrial City, 1970-1920 (Cambridge, 1983); Thomas M. Henderson, Tammany Hall and the New Immigrants: The Progressive Years (New York, 1976); David B. Tyack, The One Best System: A History of American Urban Education (Cambridge, 1976); Mel Scott, American City Planning Since 1890 (Berkeley, 1969); Dominick Cavallo, Muscles and Morals: Organized Playgrounds and Urban Reform, 1880-1920 (Philadelphia, 1981); Paul Boyer, Urban Masses and Moral Order in America, 1920-1920 (Cambridge, Mass., 1978). Students who wish to delve into the literature on urban progressivism would do well to consult the bibliographies in the following books: Ebner and Tobin, The Age of Urban Reform, pp. 156-172; Raymond A. Mohl, The New City: Urban America in the Industrial Age, 1860-1920 (Arlington Heights, IL, 1985), pp. 208-230; Howard P. Chudacoff and Judith E. Smith, The Evolution of American Urban Society, 3rd ed. (Englewood Cliffs, NJ, 1988), pp. 204-206; David R. Goldfield and Blaine A. Brownell, Urban America: A History (Boston, 1990), pp. 461-467; Bruce M. Stave, ed., Socialism and the Cities (Port Washington, NY, 1975), pp. 197-203; Buenker, et al, Progressivism, pp. 133-143. In addition, see various issues of the Journal of Urban History and John D. Buenker and Edward R. Kantowicz, eds., Historical Dictionary of the Progressive Era, 1890-1920 (New York, 1988).

48. For chapter-length syntheses, see Holli, "Urban Reform in the Progressive Era," in The Progressive Era, pp. 133-151; Mohl, "The Varieties of Urban Reform," in The New City, pp. 108-137; and Chudacoff and Smith, "The Dispositions of Urban Reform, 1880-1920," in The Evolution of American Urban Society, pp. 174-206.

CONTENTS

PREFACE

America's cities grew and multiplied rapidly during the second half of the nineteenth century. Most municipal governments were strained beyond their capacity to provide urban dwellers adequate services and efficient administration. New York, the nation's largest city, was no exception. Scores of determined New Yorkers, especially business and professional men, promoted economic facilities, attacked slums, and worked to eliminate corrupt officials. By 1900 they had provided minimum health, housing, and utility services and occasionally had elected a reform mayor.

My purpose is to describe the nature and failures of this nineteenth century reform drive and the emergence around the turn of the century of new men--progressives--with a different approach to old problems. Essentially the reformers in the nineteenth century relied on placing good men in governing positions and occasional municipal and state legislation to right wrongs. Progressives, on the other hand, worked to replace men and measures with methods of administration. In the process they partially restructured New York City's government, centralizing power in executive departments or independent commissions and giving the city continuing responsibilities for an assortment of urban ills. By 1917 an increasing number of New Yorkers were touched by an enlarged municipal bureaucracy.

Too many histories of progressivism and urban development treat as one continuous flow reform in the late nineteenth and early twentieth centuries. George Mowry in an influential book on national progressivism writes that "beginning in 1889 a succession of colorful municipal leaders gained a national reputation for their wok in urban politics." He then describes the moral crusades and work of Seth Low, Hazen S. Pingree, Samuel M. Jones, and Thomas Lofton Johnson, among others, as the essence of the new urban progressivism. Whether men such as these actually were responsible for concrete reforms is by no means clear. In a recent history of urban America Charles N. Glaab and A. Theodore Brown write "of maturing reform movements in almost every city in the country" by the 1890s and proceed to discuss reform activities on both sides of the century as parts of a single process. By focusing on one city and a few dominant issues--public administration, housing, utilities, and politics--this study attempts to discover precisely who and what groups contributed to urban reform measures and to differentiate reformers by social origin and period of activity.

I owe a debt of gratitude greater than I can express to Professor Robert H. Wiebe who supervised the entire thesis and provided needed encouragement when my perseverance was flagging. My thanks also are due my wife Dee who did all of the preliminary typing and offered comments on the entire manuscript.

Chapter 1

URBAN PROGRESSIVISM: THE NEW YORK SETTING

By the early years of the twentieth century--the progressive era--determined groups in every American city were demanding new and better housing, education, sanitation, recreation, and public utilities ranging from transportation to fresh water.[1] The city's response was complicated by the lack of an adequate political framework for efficient management and problem solving and the absence of a trained, knowledgeable leadership. Adjustment was made more difficult by the urban population. As rural Americans and foreign immigrants filled the city, the political machine, managed by a grand "sachem," won their loyalty and diligently exploited, in the interest of "boodle," the expanding functional demands on municipal government.[2]

New York City epitomized the urban crisis. If all major cities suffered under the rule of political machines, New York City had one of the most powerful in Tammany Hall; if new immigrants posed problems wherever they went, New York City received by far the largest number; if slums and dangerous working conditions blighted urban America, New York City's tenements and factories were the nation's worst; if too many women and children worked too hard and too long, New York City employed the greatest number; and if businessmen everywhere purchased franchises and evaded taxes, New York City sold the richest urban prizes in the land.

City government, as Lord Bryce observed, was America's most conspicuous failure, and a great diversity of reform groups arose to correct it.[3] In fact, these ambitious urban progressives early in the twentieth century came to shape much of the state and even national reforms in such areas as child labor, tenement and factory legislation, and public utility regulation. Therefore, an analysis of the components of reform in New York City illuminates the broader process during the progressive era.

In order to avoid a priori judgments of who was or was not a progressive, this analysis begins not with men but with the reform issues themselves. These fell into three general clusters. One group centered around modifications in the framework and administration of New York City's government. Here progressives desired a more honest and efficient government, enlarged powers of home rule, the rationalization and centralization of executive administration, fairer and more direct nominating and election procedures, and an expansion of civil service to eliminate partisanship in the choice of city employees. These changes, it was assumed, would create a responsive government of service at minimum cost and waste.

A second group of issues focused on the expansion and regulation of the city's economic facilities. Debate revolved around private versus public ownership, construction, and operation of utilities; the city's right to control franchise grants; and municipal regulation of privately owned urban services. A third group concerned the quality of individual life in the urban environment, particularly the disadvantaged individual. This area included such issues as the reform of tenement-house and factory conditions, and the expansion of welfare services, hospitals, public lavatories and baths, parks, playgrounds, and schools.

Of course the problems in these three categories were often interrelated. The drive for more efficient municipal administration was paralleled by a similar concern about the operation and regulation of public utilities. Demands for government regulation of utility rates were comparable to demands for the regulation of factory conditions. In fact, the simultaneous development of the Bureau of Municipal Research, public service commission, Board of Water Supply, and tenement house commission in New York City reveals a common core of concerns and motives among a broad range of reformers.

A multitude of civic, business, professional, and political organizations institutionalized the reform drive.[4] They provided reformers a means, alternative to regular party channels, through which to appeal to the public, launch election campaigns, criticize officials, and advocate reform legislation. Organizations with large memberships, full-time staffs, and interests in a wide range of city affairs exerted a continuing influence on municipal problems. Other permanent organizations with more narrow interests concentrated on specific issues of city-wide importance, or issues local or geographical in orientation. Still others, ad hoc in nature and eclectic in membership, were formed to do battle at a specific time over a particular issue such as a vice crusade or a municipal election.

Beneath the crust of organization lay a great diversity in reform types. At least five can be distinguished.[5] A number of well-to-do gentlemen, generally self-employed or professional and from families of reputation, devoted much time to civic activities. Since the 1870s, these mugwumps, as they often were labeled, made working for cleaner government an avocation. Yet they brought no specific occupational skills to bear on city problems. Instead they hoped to secure a more responsive government through the election of honorable men, invariably men like themselves.

Other reformers, younger, middle-class in origin and possessing unique abilities in professional fields such as medicine, law, administration, engineering, and social work or in business and labor, consciously sought to apply their knowledge to curing New York's ills. Confident that their unique skills were important in an increasingly urban-industrial America, these progressives sold their expertise where it was accepted and moved in and out of municipal departments and reform organizations.

Closely related to this second regiment of reformers were the academicians, usually professors of political science, economics, or sociology in the nation's leading colleges and universities. Seeing the city as a ready-made laboratory, they sought to apply textbook theories on municipal government and its functions to New York's problems; and city needs often served as the basis for curriculum development.

Swelling the reform ranks was a fourth group comprised of individuals united mainly by a broad humanitarian or Christian concern for the discontented in urban life. Included were journalists who exposed evil and corruption, an assortment of clergymen who preached environmental rather than personal salvation, and college-educated women who were appalled at slum living, especially its effects on women and children. They established settlement houses, attacked vice and gambling, campaigned for better working and living conditions, worked in elections, and lobbied for legislation. Legislators, mayors, governors, and other politicians with reform leanings who were instrumental in getting progressive legislation enacted filled out the reform camp. All five reform types were active throughout the progressive years and often worked together within a single organization or for some specific end. Shared activities perhaps blurred but by no means eliminated their distinct responses to New York's problems.

NOTES

1. The following books have been most helpful in the formulation of the ideas and organization of this chapter. Samuel P. Hays, The Response to Industrialism: 1885-1914 (Chicago, 1957); Richard Hofstadter, The Age of Reform (New York, 1955); Barry Dean Karl, Executive Reorganization and Reform in the New Deal (Cambridge, Mass., 1963); Theodore J. Lowi, At the Pleasure of the Mayor: Patronage and Power in New York City, 1898-1958 (London, 1964); George E. Mowry, The Era of Theodore Roosevelt, 1900-1912 (New York, 1958); Robert H. Wiebe, Businessmen and Reform: A Study of the Progressive Movement (Cambridge, Mass., 1962), and The Search for Order, 1877-1920 (New York, 1967).

2. Constance McLaughlin Green, The Rise of Urban America (New York, 1965), pp. 75-111; Clifford W. Patton, The Battle for Municipal Reform (Washington, D.C., 1940), pp. 8-18; Arthur M. Schlesinger, The Rise of the City 1878-1898 (New York, 1933), pp. 73-74; Karl, Executive Reorganization, p. 91; Horace E. Deming, "The Municipal Problem in the United States," in A Municipal Program: Report of a Committee of the National Municipal League (New York, 1900), pp. 38-54.

3. Harold U. Faulkner, Politics, Reform and Expansion (New York, 1959), p. 23.

4. Lowi, Pleasure of the Mayor, pp. 103, 107, 190, 201.

5. These are suggested in Hays, Response to Industrialism, pp. 73-75; Karl, Executive Reorganization, pp. ix, x; Lowi, Pleasure of the Mayor, pp. 52, 66-67; Wiebe, Search for Order, pp. 111-132, 164-176.

PART I

THE NINETEENTH CENTURY REFORM TRADITION

New York's urban patricians, who reached maturity as reformers in the last decades of the nineteenth century, experienced the industrial and ethnic upheavals largely as spectators. Their wealth and place in society were inherited and therefore unrelated to the newer forms of business or professional activity. Rooted in New York, accustomed to deference, if not power, and bewildered by the passing of a city they called home, these men of manners before the turn of the century naturally took the lead in singling out enemies and directing reform attacks.

Chapter 2

THE POLITICS OF REFORM

For New York's men of importance, Tammany Hall, the Democratic organization that governed the city, was the ultimate source of the many ills that plagued Gotham. Efforts to dislodge Tammany from power easily took on the flavor of crusades against sin. Identifying with the forces of good, New York's better citizens broadened their appeal by underplaying party labels and stressing the virtues of nonpartisan local elections. For tactical reasons the reformers occasionally subordinated ideology and sought alliances with independent Democrats, the city's few Republicans, and other politically unaligned groups.

In the 1870s a nonpartisan Committee of Seventy, comprised primarily of businessmen, had successfully dislodged from City Hall the notorious Tweed Ring. The committee had been organized "to rescue New York from the slough of bankruptcy and disgrace into which it had fallen under the rule of knavish partisans."[1] Unfortunately, the rescue was short-lived; a successful skirmish against Tweed did not mean the death of the Tammany Tiger. In the decades after 1870 New York continued to be intermittently subjected to machine rule. Reformers responded with various clubs, committees, and independent candidates. Their hopes for a permanent revolution mounted sharply in 1894 when William L. Strong, a reform candidate, was elected mayor.

The catalyst for that triumph was supplied by the Reverend Charles W. Parkhurst, president of the Society for the Suppression of Crime and of the City Vigilance League. Born in Framingham, Massachusetts, and a graduate of Amherst College, Parkhurst in 1880 accepted the pastorate of the

Madison Square Presbyterian Church, one of New York's largest and most influential congregations. Shocked by the slums, this energetic pastor plunged into charity, education, and missionary work and at the same time cried out against Tammany's evil influences, especially the relationship between commercialized vice and police corruption. In disguise, he and several detectives collected evidence of the Police Department's complicity in extortion, gambling, and violations of the liquor laws.[2]

In January, 1894, Harper's Weekly pointed out that Parkhurst's cause was in fact "a war of the citizens of New York against Richard Croker and his kind, against his hall, and against the wretched system of municipal government which has made blighting things like him and his possible." New York, it predicted, was "at the beginning of a civic revolt against criminal municipal government, not merely of an effort to secure more virtuous police captains and inspectors."[3]

The Chamber of Commerce was spearheading that revolt. Prodded by President Charles Stewart Smith and several others, the Chamber called for a state investigation of the city's police department. The New York State Legislature then appointed such a committee under the chairmanship of Clarence Lexow, and when the Democratic Governor Roswell P. Flower refused to sign the necessary appropriation bill in 1895, members of the Chamber guaranteed the Lexow committee sufficient funds to proceed. In addition, the Chamber recommended John W. Goff as legal advisor and aided in gathering data and securing witnesses.

Parkhurst's dramatic revelations and the facts accumulated by the Lexow committee brought back memories of the Tweed days. The investigation revealed that Tammany Hall controlled the police force and exacted tribute from police officers seeking promotions. For the benefit of the Hall's coffers the police collected protection money from legitimate businessmen and blackmailed prostitutes, gamblers, and other criminals. Under police cover Tammany's men stuffed ballot boxes, threatened poll watchers, and stole elections.[4] Reformers such as Anson Phelps Stokes, a merchant and banker interested in civic betterment, urged all men of good character and successful business experience to unite and District Court Justice John Jeroloman demanded an organization that would "drive those who have violated the laws of our Commonwealth from power." The most important and delicate task for those initiating the drive against Tammany was to coordinate the heterogeneous group of reform organizations, Republicans, and independent Democrats into one political movement.[5]

The German-American Reform Union, organized by Gustav H. Schwab, shipping merchant and prominent member of the Chamber of Commerce, made an unsuccessful attempt to rally supporters in September, 1894. The initiative then passed to the Chamber itself, and a group of its members who were also active reformers called a meeting on the sixth of September at Madison Square Garden. From that meeting came a Committee of Seventy, composed of "grave and pompous citizens--chief men of the town," to confer with other organizations and direct a general reform campaign. The prominent lawyer Joseph Larocque and Charles Stewart Smith were its leading officers and such men as Cornelius Vanderbilt and J. Pierpont Morgan financed the committee. After many conferences with various political groups, the Committee of Seventy decided to nominate William L. Strong, one of its own members for mayor, and even the Republicans agreed to support him.[6]

The reformers had chosen a representative man. Born in Ohio, Strong came to New York in 1853 and entered a dry goods firm. In the tradition of Horatio Alger, he soon formed his own successful company and subsequently became president of the Central National Bank and the Hamer Lee-Bank-Note Company, vice-president of the New York Security and Trust Company, and a director of such companies as the Erie Railroad and New York Life Insurance.[7] To those who assumed that success in business guaranteed success as a political leader,

Strong was well qualified to "throw out the rascally politicians and give . . . a good business administration."[8] That, in fact, was his creed. "No one who has had experience in the administration of city affairs will contend that it is other than a business proposition," he later affirmed. "That is all there is to it."[9] That fall his platform, which was drafted by the Committee of Seventy, emphasized nonpartisan government and such improvements as additional schools and public baths and cleaner streets.

Strong defeated Tammany's candidate, Hugh J. Grant, by 42,627 votes. The Committee of Seventy, confident that it had struck a decisive blow for better government, disbanded early in 1895, leaving the new mayor shorn of the continuous support traditionally supplied by political parties.[10] With no permanent reform party to woo them, thousands of New Yorkers by the close of Strong's administration in 1897 had gravitated back to Tammany "as an inveterate drunkered [sic] returns to his liquor."[11] The reformers found their solace in the formation of a new independent citizens' movement.

In December, 1896, a group of leading New Yorkers, many of whom were members of the Chamber of Commerce, the City Club, and other civic and business groups active in reform causes since the 1870s, met in the law office of Elihu Root to plan the new citizens' organization. A sense of urgency pervaded this and subsequent meetings. For the first time in New York's history the issues in the election of 1897 would be purely local. Three years before an amended state constitution had separated municipal from all other elections, and this was to be the first election for the newly consolidated Greater New York. Not only would Manhattan and a slice of the Bronx need to be kept out of the Tiger's claws, but Brooklyn, Queens, Richmond, and the entire Bronx as well.

Under Root's guidance, a committee of these gentlemen-reformers, after conferring with several religious, civic, and labor leaders, announced early in January, 1897, the formation of the Citizens' Union as a municipal political party.[12] The Citizens' Union brought the political exclusiveness and evangelistic fervor of the crusader to the battle for mayor. "The election of 1897," its first announcement read, "will be a crisis so important as to demand of every citizen the full performance of his public duty."[13] Albert Shaw, expert on municipal problems and editor of the Review of Reviews, wrote, "I have never known a time in the history of any great city . . . when so much was at stake for the cause of good municipal government at one single election as will be at stake in this city this year."[14]

Fusion with the Republicans, so successful in 1894, seemed immoral in 1897. In a letter to Low, Richard Watson Gilder, magazine editor, tenement reformer, and member of the Union's executive committee, called the campaign "something in the nature of a revolution. The mass of the people recognize that there is a cause; and only a cause can win this time--not an old fashioned combination."[15] Many independent citizens, moreover, felt betrayed by the Strong administration's apparent capitulation to Republican demands. "The Citizens' Union was organized to fight against the Platt machine's influence in the city's politics just as strongly as against Tammany's influence," declared John Brooks Leavitt, at a mass meeting in July. "It cannot make a bargain with Mr. Platt without sacrificing its principles."[16] "We want no Platt-and-Croker deals, we've had enough of them" ran one of the campaign songs.[17]

The Citizens' Union's only alternative was to nominate its own candidate before the Republican convention met, then hope all others would endorse its man. At the end of May, the Union asked Seth Low to accept that nomination. Like Strong, Low epitomized the independent, wealthy citizen who dominated reform in the 1890s. His father, Abiel Abbot Low, had moved in 1829 from Salem to Brooklyn, entered the China trade, and founded the tea and silk house bearing the family name. The young Seth had all the advantages of wealth: a house on fashionable Brooklyn Heights, summers in New England, travel in Europe, and education at Columbia College and a year at Columbia's law school. Soon after joining his father's firm in 1875, Seth Low abandoned an active business career to work for civic improvement. He served two terms as mayor of Brooklyn and was a member of the rapid transit and Greater New York charter commissions. When asked to run for mayor he was serving as president of Columbia University.[18]

The leaders of the Citizens' Union put great pressure on Low to accept the nomination because they feared his refusal would endanger the whole good government movement. Charles Stewart Smith assured Low that his candidacy was necessary if the Union were to gain German, Republican, and anti-Tammany Democratic support as well as the approval of the major newspapers. Others in the Union echoed this sentiment and begged Low to accept because of New York's influence in the nation. "Your consent to be a candidate would have an enormous moral influence for good and disinterested city government throughout the whole union," wrote Albert Shaw. "It seems to me you could, as mayor for the next four years give an impulse to the right development of Greater New York which would be felt for half a century."[19]

Low, a Republican in national politics, was unconvinced that the Citizens' Union should go it alone. He feared that hasty action would alienate the local Republican organization and make a strong anti-Tammany coalition impossible. Therefore he asked his supporters for further assurances that his candidacy would unite the friends of good government and receive popular support. In deference to Low's wishes, Charles Stewart Smith persuaded the executive committee of the Citizens' Union to delay for several months the tendering of a formal nomination.

The reformers quickly gathered signatures on a petition begging Low to run for mayor, and the Citizens' Union established a network of district organizations to extend its influence. Their efforts were bolstered when a group of gentlemen in Brooklyn formed a Committee of Fifty to encourage Low's candidacy and seek Republican support. Although the committee failed to rally the Republicans behind Low, the Citizens'

Union, supported by its Brooklyn friends, still tendered Low the nomination on September 1 and destroyed all chances for a genuine fusion movement.[20] Principle had triumphed over compromise.

Low accepted the nomination. "I could not have declined to accept the nomination of the Citizens' Union, under the circumstances, without throwing everything into chaos and making the election of a tammany candidate practically certain," Low wrote his friend Theodore Roosevelt. "Besides this, there would have been the destruction of one of the most remarkable efforts of the civic spirit witnessed in our day in the City of New York."[21] Still, Low continued to hope for a Republican endorsement. But at the Republican nominating convention on September 28, Lemuel E. Quigg, president of the Republican county committee, convinced the delegates the Citizens' Union wanted "to smash the Republican Party" and maneuvered the nomination of Benjamin F. Tracy for mayor. Neither Low, Tracy, nor Henry George, the single taxer and candidate of still another anti-machine group, could defeat Robert A. Van Wyck, chief justice of the city court of New York and Tammany's man. The reformers took heart in the news that Van Wyck's total vote was less than that of the other three candidates combined.[22]

The Van Wyck administration reflected Tammany at its worst. "It was well known that everything obtained by favor from the Van Wyck administration was bought and paid for," reminisced one New Yorker.[23] A legislative investigation revealed that New York was "at the mercy of a gang of men, of whom Croker is the chief, and who constitute the controlling power in the Tammany organization, which, in turn, controls the municipal government."[24] Graft and tribute exacted from saloonkeepers, gamblers, and prostitutes oiled the Tammany machine; municipal employees regularly blackmailed businessmen dealing with the city.[25] It was The Old, Old Story," lamented Harper's Weekly. News of official looting, vice, and depravity "have become as truly a part of our breakfast-table adornment as the coffee-urn itself."[26]

The Citizens' Union was determined to write "a new story." Instead of disbanding after its failure to elect Low in 1897 the Union reorganized as a municipal political party. Having learned "that it could not unaided expect to terminate Tammany control," the Citizens' Union's central city committee early in 1900 began preparing for a fusion movement the following year.[27] No effort will be spared "to secure the active co-operation of all organizations, societies and individuals qualified to render effective service," promised the Union's first election manifesto.[28] Anti-Tammany sentiment was contagious in 1901. The City Club voted to pursue a united front. Diverse groups such as the Merchants' Association, the Franchise Tax and Municipal Ownership League, the Manhattan and Brooklyn Single Tax Clubs, the Reform Club, the German-American League, the City Vigilance League, and the Good Government Clubs, joined the crusade. A number of independent Democratic groups and the Republican party rounded out the coalition. The task of effectively uniting these groups behind a single program and slate of candidates largely fell to the Citizens' Union and Republican party.

On April 25, 1901, the Citizens' Union appointed a committee of one hundred, headed by its own chairman, Robert Fulton Cutting, philanthropist, president of the Association for Improving the Condition of the Poor, student of municipal government, and scion of an old-line banking family. A platform, labeled socialistic by some New York newspapers, calling for more home rule, public baths, parks, schools, and utilities, and more effective city control of franchises and tenements was adopted. The committee sponsored several anti-Tammany conferences attended by representatives from a host of clubs and political organizations. Besides the Citizens' Union the more influential included Republican county organizations, independent Democratic groups, the German-American Reform Union, the City Club, the City Vigilance League, and the

Merchants' Association.[29] They were united only in their desire to overthrow the existing reign of "political corruption, extravagance and indecency."[30]

Twelve influential members of the Citizens' Union's committee negotiated during August and September with the various anti-Tammany groups and prepared a list of possible mayoral candidates. Of the four leading nominees--John DeWitt Warner, George F. Peabody, Seth Low, and George L. Rives--the Citizens' Union leaders favored Low. On September 18, 1901, a full conference of anti-Tammany organizations voted to battle Croker's Tiger behind Low. Tammany's Richard Croker shrewdly convinced the Democrats to nominate as Low's opponent an honest Brooklyn lawyer and reformer Edward M. Shepard.[31]

Low defeated Shepard by over thirty thousand votes, but after two years of relatively scandal-free and unexciting municipal government, most New Yorkers were disillusioned with the "pompous righteousness" and cautious timidity of the reform administration. Many of the district leaders of the Citizens' Union urged the central committee not to renominate Low in 1903. Robert Fulton Cutting, Carl Schurz, Reverend Charles W. Parkhurst, and Everett P. Wheeler, supporters of Low and influential in the higher councils of the Union, disregarded these pleas and convinced the central committee again to support Low as a fusion candidate.[32]

On September 9, Low was nominated by a conference of representatives from the Citizens' Union, the Republican party, and various ethnic organizations. The broader-based civic and commercial organizations had no part in the nominating procedures. Many independent Democrats, especially John C. Sheehan's Greater New York Democracy, opposed Low and in the course of the campaign gravitated toward the Tammany camp. Charles F. Murphy, the new pragmatic boss of the Wigwam, further confused fusion sentiment when he engineered the nomination of George B. McClellan, the charming, scholarly, and honest son of the Civil War general, and then had the

Democrats endorse the fusion candidates for comptroller and president of the Board of Aldermen, Edward M. Grant and Charles V. Fornes. Embarrassed but principled, the fusion leaders withdrew the nomination of its two candidates and selected others to run with Low. The popular McClellan carried four of the five boroughs and won by over 61,000 votes.[33]

Low's brief administration marked a watershed. The leadership in the electoral campaigns that culminated in Low's victory consisted of men whose interests, assumptions, and abilities properly belonged to the late nineteenth-century revolt of New York's better citizens against corrupt urban government. Observation reinforced by logic informed their perception of urban problems. Ultimately, a number of them believed, all municipal ills were rooted in men's behavior. Corruption, waste, and other evils, whether in one's private or public world, inevitably resulted from unethical or immoral conduct. In fact, for the mugwump who dominated reform politics in the nineties, the distinction between public and private morality was blurred. Rules of conduct, rooted in absolute laws of right and wrong, related to individuals, not spheres of activity. New York's mismanaged and unresponsive government was simply the logical doing of dishonest and bad men. "Are we a community of honest men, or is our city to be forever branded as a den of iniquity?" was a question New York's better citizens could understand.[34]

Bad men were not difficult to locate and describe. Politicians running the city on Tammany's behalf were the most visible. More powerful but elusive were the bosses operating out of neighborhood saloons and clubs and owing fealty only to the Grand Sachem who from his headquarters on Fourteenth Street managed the Democratic organization, hand-picked candidates, dispensed favors, adjusted conflicting claims, disciplined the erring, and ruled New York as "a general commanding an army."[35] A "species of monarchs," Harper's Weekly called Tammany's leaders; knaves, plunderers, bandits, thieves, and spoilsmen were less exalted but more commonly used labels. Belonging to the cast of bad men, but in subordinate roles, were gamblers, saloonkeepers, poolhall operators, keepers of houses of ill repute, and the like--"the interests of banded vice"--upon whom Tammany exacted tribute. New York's lower classes, especially the foreign born, were regarded as captives rather than subjects of the enemy. Pitied more than despised, considered ignorant rather than bad, they were the souls for whom the reformers battled the bosses.[36]

"United good men" must remove from office the "combined bad men," pleaded Harper's Weekly in 1893.[37] New York's patrician knew to which group he belonged, and the inverse of his own logic seemed to dictate such a strategy which, by the decade of the nineties, had become part of an orthodox reform tradition going back at least to the successful overthrow of Tweed. Only with new men in office, declared a Good Government Club brochure in 1894, would properly cleaned and policed streets, pure water, and equitable tax rates be secured.[38] Understandably, then, political activity absorbed much of the reformers' energies. Small in number and with no influence in the local Democratic or Republican parties, these leading citizens formed political committees and civic clubs and used their professional and business organizations to unite their efforts, mobilize financial support, and establish community-recognized bases from which to seek popular support. Under this organizational umbrella reformers shuffled from one group to another and often were active in several simultaneously.

Most of the leading activists were professional men, generally lawyers, and businessmen. Various editors, journalists, ministers, and an assortment of settlement house people moved in and out of the reform activities. In 1894 the Committee of Seventy, which engineered Mayor Strong's election, reflected well this constituency and believed it was continuing in the tradition of its namesake, the committee that overthrew Tweed.[39] Three years later a number of these same gentlemen--Robert Fulton Cutting, Charles Stewart Smith, J. Kennedy

Tod, William B. Hornblower, John Claflin, Jacob H. Schiff, and Gustav H. Schwab, among others--all either lawyers or businessmen and all leading members of the Chamber of Commerce, City Club and similar organizations, put together the Citizens' Union.[40]

The formation of the Citizens' Union in 1897 as a municipal political party represented no break with the past. Rather it institutionalized a practice only loosely and impermanently worked out by a short-lived Peoples' Municipal League, which ran candidates in 1890, and the Committees of Seventy. Other reforms popular during the latter decades of the nineteenth century, such as the separation of municipal from national elections, home rule, primary and ballot reform, the expanded use of civil service, and increased power of appointment and removal for the mayor, were meant to ease the way for men of good character to gain control of government and minimize interference with their subsequent enforcement of honest, clean, and progressive government in New York.[41]

To choose the right tactics at the proper time was the most vexing and divisive problem faced by the reformers. They frequently sought alliances with other anti-Tammany elements, most notably independent Democrats and Republicans. Such coalitions gave a multipartisan--or by the tortured logic of the reformers, a nonpartisan--cast to their efforts and provided a practical means of outflanking Tammany. By joining with others the reformers imagined they were participating in popular uprisings. Their own rhetoric nurtured the illusion. "Our city's for the people," ran the Citizens' Union campaign song; the struggle for good government was one "between the people and the politicians," "between good citizens and Tammany hall."[42] When the City Club was formed in 1892, Harper's Weekly described it as an "organization of public spirit."[43] The municipal political party organized in 1897 was a "citizens' union." Reform choices, of course, were "citizens' candidates."

The fragility of these fusion movements made tenuous

the best-laid plans. In 1894 the Republicans refused to join with the Committee of Seventy until a candidate of their liking was selected. When the Citizens' Union insisted on nominating a candidate for mayor in 1897 without prior approval of the Republicans, Platt's party refused to join the coalition. The various independent Democratic groups, never completely out of the shadow of the dominant organization, were even more unpredictable. Low and the Citizens' Union discovered this in 1903. While Tammany more often than not served as an integrating symbol for the anti-machine elements, it proved divisive as well. The Hall's nomination of Edward M. Shepard for mayor in 1901 kept many reform-leaning Democrats in line. That "Mr. Shepard's good character, in winning votes, is . . . a direct menace to the cause of good government in this city" was recognized by Harper's Weekly.[44] The astute Charles Murphy not only ran McClellan for mayor in 1903 but threw the reformers in disarray by endorsing two of their candidates.

The reform camp itself tended to divide on tactics and on the meaning of a nonpartisan ticket. "The Mugwump element is without discipline and cannot be got together," complained Edmond Kelly, the energetic lawyer who was largely responsible for organizing the City Club in 1892.[45] The year after Strong was elected, the "rash young Daniels" of the Good Government Clubs, refused to back a slate of candidates for state and local offices fielded by a coalition of Republicans, Independents, and a committee of fifty formed by the Chamber of Commerce.[46] Believing the ticket was the result of horse trading among the fusionists and thus serving partisan interests, these goo-goos ran their own candidates and righteously declared that their self-respect and the confidence of the community could not be jeopardized "even for the sake of defeating Tammany Hall."[47] When the Citizens' Union trod the same path in 1897 and decided principle was more important than compromise in selecting a mayoral candidate, Richard W. G. Welling, the young lawyer who organized many of the Good Government Clubs, caught well the irony when in his diary he expressed amazement that the older men running the Union were then pursuing a course that just two years before they considered rash. Candidate Low feared the Union's intransigence was political suicide. In fact, the man who carried the reformers' banner in three campaigns was less than enthusiastic about the whole idea of a municipal political party.[48]

Until the Citizens' Union was formed, no organization kept the reform element united after an election. Even the Union was a weak reed as many of the gentlemen-reformers divided their time among several groups and clubs and quickly turned to their private interests. Unlike the Tammany man, the reformer's main business was not politics. "From time to time, when the conduct of municipal affairs by Tammany Hall had become conspicuously outrageous," explained the reformer Carl Schurz, "a committee of public-spirited gentlemen would, so to speak, descend from the clouds to appeal to the conscience and enlightened patriotism of the community . . . and then, when the fight was over, to disappear again from the scene. . . ."[49] Once a campaign ended, the larger anti-Tammany coalition lost its reason for being and likewise disintegrated. "These reform movements are like queen-hornets," an East Side gangster remarked, "they sting you once and then they die."[50] "Mornin' glories" was how George Washington Plunkitt, the colorful and perceptive Tammany boss, described the movements. They "looked lovely in the mornin' and withered up in a short time."[51] Wholly accurate or not, these metaphors captured the frustration of mugwump politics.

Even victory did not alter the pattern of discontinuity. After Strong's election the Committee of Seventy disbanded. The new mayor complained that many reformers, very free with advice, refused as a matter of principle to accept positions in his administration.[52] Several Citizens' Union men, after agreeing to serve under Low in 1901, felt compelled as a concession to "a high ideal of service" to sever relations with the Citizens' Union and thus at a crucial time sapped the strength of that organization.[53] When Strong and Low sought to build and maintain support by rewarding political supporters with jobs, they were criticized by many of the reformers, especially the Good Government Club leaders. The Citizens' Union bemoaned the fact that one of its district organizations demanded patronage from Low. Both reform administrations were severely criticized, especially by independent Democrats, for catering to Republican demands. That both mayors were Republicans in national politics, and openly paraded their labels, further irritated the non-Republican elements in reform but did not spare Low or Strong from the Republican charge of being denied its fair share of patronage. To satisfy all the anti-Tammany groups was impossible; to satisfy some meant excluding others inevitably destroying the coalition.[54]

Criticizing the established parties and eschewing their methods, the reformers built no alternative structures through which one of their own could exercise power in a continuous and harmonious way. The reformers never could understand Plunkitt's dictum: "Politics is as much a regular business as the grocery or dry-goods or the drug business. You've got to be trained up to it or you're sure to fail."[55] Both mayors virtually were abandoned in their task of carrying out a reform program. Other than through the force of their personalities, Low and Strong had no means to compromise and balance the conflicting demands upon their administrations. Unfortunately, neither man could inspire confidence, exercise discipline, hold friends, convert foes, or establish rapport with the mass of lower-class voters.[56]

More fundamental was the limitation of mugwump strategy. The act of placing upright men in office was assumed to be self-fulfilling. The reformers' rhetoric at least implied that much. The people's city that they talked and sang about was "to be ruled by honest men." Just how these men were to govern--that is, how they were to extend a morality and ethics essentially personal to the conducting of the complex affairs of New York's government--was rarely considered. Because they interpreted the entrenched reign of "political corruption, extravagance and indecency" primarily as a function of sinful personality, mugwumps concentrated on changing the governing personalities.[57]

So sure of their own uprightness, reformers naively assumed their men in office would not exhibit the foibles common to all men. The Strong and Low administrations were free from the worst machine abuses. But they still faced the gritty details of patronage, dealing with special interest groups, cajoling administrative heads to run their departments more efficiently, coordinating conflicting demands of reform groups, working with recalcitrant legislatures, and establishing rapport with the people. Here in the real world of politics and power, disagreements easily arose, personalities clashed, friends became enemies, and the most well-intentioned plans and goals stalled.

While Strong was accused by some of moving too slowly in removing Tammany holdovers, Low was severely criticized by his supporters for not removing quickly enough his own incompetent appointees. Both mayors were accused of doing too little to increase the city's parks, schools, playgrounds, public baths, and other socially desirable facilities. In 1897 Welling in his diary confided that "Strong's 'cant & hypocrisy' disgusts everyone with sham reform."[58] Seven years later a settlement house worker described reform under Low as "conservative and autocratic"; administration leaders, he wrote, catered to men of property and "saw to it that that powerful ten percent of the people that pay taxes should in no case find grounds for complaint against 'popular expenditures.'"[59] For many, enthusiasm easily turned to disillusionment.

The reform administrations were not without achievements. Positive programs were pushed, especially by the Citizens' Union. Under Strong and Low civil service rules and tenement laws were more honestly enforced, a few new parks and schools were built, improvements were made in the city's water and transportation systems and better men were appointed to head city departments, which were run more efficiently. Low even opened all municipal board meetings to the press, held weekly meetings with department heads, and made some minor changes in departmental procedures. The Citizens' Union cheered and campaigned to extend these limited achievements. The Union's literature also called for more strict supervision of utilities, especially franchise grants, municipal ownership of water supply and if necessary other utility services, enforcement of the eight-hour day for all city employees, and the payment of prevailing wage rates.[60]

Too infrequently asked or even considered, however, were questions that probed beyond particular panaceas. Were existing administrative methods adequate to regulate tenements or the civil service? Did the rapid transit board possess sufficient power and knowledge to provide needed transportation? Was the city's bureaucracy sufficiently developed to efficiently regulate and operate utilities? Were finances, plans, and methods for extending welfare services and public projects available? Did the city know what type of schools it wanted and possess the administrative means and expertise to manage a greatly enlarged system? Were criteria available to evaluate the need and skills required for various city jobs? In his quest to purify and make more responsive New York's government, the mugwump was concerned more with men than with means.

To whom the government was to be responsible was unclear. Rhetoric only obscured the realities of the social and economic divisions within New York. The reformers, all men of means, campaigned in the name of a people they neither knew nor understood. It was ironical that men who analyzed government as a function of human behavior failed to comprehend the behavior of millions of poor New Yorkers. For scores of municipal employees efficiency drives in city departments often meant fewer jobs, reduced salaries, and longer working hours. The tenement dweller received scant comfort from the knowledge that the Strong and Low administrations computed property taxes more accurately or eliminated graft in contract awards. The foreign-born could hardly become enthusiastic about a government that more effectively apprehended unlicensed peddlers. Lower tax rates, better civil service enforcement, more carefully regulated franchise grants and utility services, home rule, shorter ballots, better garbage collection, and the like--all lumped together and labeled a business administration--meant too little in a personal way for thousands of disadvantaged Gothamites. They needed bail and counsel when in trouble with the law, a free bucket of coal when penniless, and to be "put on a broom" when physically unfit for hard work.[61]

Tammany proved a more perceptive judge of human nature and the immediate needs of New York's lower classes. "I just study human nature and act accordin'," stated Plunkitt in his delightful Very Plain Talks on Very Practical Politics. "To learn real human nature you have to go among the people, see them and be seen," he explained. Not political arguments or literature, but political philanthropy was what the poor understood and appreciated. "If a family is burned out I don't ask whether they are Republicans or Democrats, and I don't refer them to the Charity Organization Society, which would investigate their case in a month or two and decide they were worthy of help about the time they are dead from starvation," boasted the sagacious Plunkitt. "I just get quarters for them, buy clothes for them if their clothes were burned up, and fix them up till they get things runnin' again." Not only were the poor "the most grateful people in the world," continued Plunkitt, "they have more friends in their neighborhoods than the rich have in theirs."[62] Too many reformers never seemed to learn that lesson. One of their own, the young lawyer Richard Welling, expressed his despair by quoting from the Book of Job: "The Tabernacles of the robbers prospered/and they that provoked god were secure."[63] What Welling did not anticipate was that the "tabernacles" would be assaulted in other than frontal attacks, and with new weapons.

NOTES

1. Citizens' Union, Campaign Book of the Citizens' Union, 1897 (New York [1897]), p. 23; M. R. Werner, Tammany Hall (Garden City, N.Y., 1928), p. 222.

2. George Francis Knerr, "The Mayoral Administration of William L. Strong, New York City, 1895-1897" (unpublished Ph.D. dissertation, New York University, 1957), pp. 10-11; William J. Schieffelin, "Reminiscences," Columbia University Oral History Collection (hereafter cited CUOHC), pp. 16-17; "Do Criminals Govern the Police?" (editorial), Harper's Weekly, XXXVIII (January 6, 1894), 3; "The Rev. Dr. Charles Parkhurst," ibid., March 31, 1894, pp. 299-300; New York World, March 1, 1892.

3. "Police Department," XXXVIII (January 13, 1894), 40.

4. New York World, January 19, 1894; Annual Report of the Corporation of the Chamber of Commerce of the State of New York, 1893-1894, pp. 77-78, 1894-1895, p. 42 (hereafter cited CC Annual Report); Knerr, "Mayoral Administration of Strong," pp. 18-19; "Report of the Special Committee Appointed To Investigate the Police Department of the City of New York," New York State Senate Documents, No. 25, III (118 Sess.), 1895, pp. 3-11, 15-51 (hereafter cited "Committee to Investigate the Police Department").

5. John Jeroloman to Charles S. Fairchild, November 13, 1893, Papers of Charles S. Fairchild, New York Historical Society Library (hereafter cited NYHSL); Anson Phelps Stokes to Fairchild, November 9, 1893, ibid.; "Committee to Investigate the Police Department," pp. 16-58; "The Necessity of Anti-Tammany Union" (editorial), Harper's Weekly, XXXVIII (June 9, 1894), 531; "The People and the Politicians" (editorial), ibid., XXXVIII (July 28, 1894), 699.

6. Richard W. G. Welling, "City Reform Club," Personal Misc., Folder C, Papers of Richard W. G. Welling, New York Public Library (hereafter cited NYPL); Knerr, "Mayoral Administration of Strong," p. 28; "Germans and City Reform" (editorial), Harper's Weekly, XXXVIII (June 30, 1894), 603; Catalogue of Portraits in the Chamber of Commerce of the State of New York (New York, 1924), p. 91; "Committee of Seventy" (editorial), Harper's Weekly, XXXVIII (October 13, 1894), 966; A. C. Bernheim, "The Committee of Seventy," in The Triumph of Reform: A History of the Great Political Revolution, November Sixth, 1894 (New York, 1895), pp. 7-14. Many members of the Committee of Seventy played important roles either in financing or working in the many reform organizations and political campaigns throughout the 1890s, such as L. J. Callanan, John Claflin, Lewis L. Dalafield, T. A. Fulton, Abram S. Hewitt, William B. Hornblower, Dr. A. Jacobi, W. Travers Jerome, Woodbury Langdon, Joseph Larocque, J. Pierpont Morgan, James B. Reynolds, G. L. Rives, W. Harris Roome, William J. Schieffelin, Jacob H. Schiff, Gustav H. Schwab, E. R. A. Seligman, Charles Stewart Smith, Simon Sterne, William L. Strong, J. Kennedy Tod, Cornelius Vanderbilt, and Everett P. Wheeler.

7. "William L. Strong," Harper's Weekly, XXXVIII (November 17, 1894), 1099.

8. Lincoln Steffens, The Autobiography of Lincoln Steffens (New York, 1931), p. 255.

9. Quoted in Citizens' Union, Campaign Book, 1897, p. 34.

10. Ibid.; New York Times, November 7, 1894; Knerr, "Mayoral Administration of Strong," pp. 40, 273-275.

11. Werner, Tammany Hall, p. 445.

12. Walter T. Arndt, "A Quarter Century of the Citizens' Union: A Chapter in the Political History of New York City," Searchlight, XII (December 14, 1922), 3-5; Citizens' Union, Printed Circular of Citizens' Union Committee of Organization, April 9, 1897, Campaigns, Box H-1, Citizens' Union Collection,

Columbia University (hereafter cited CU Coll.); Lawrence A. Tanzer, "Reminiscences," CUOHC, p. 19; Elihu Root to Fairchild, February 29, 1896, December 9, 1896, Fairchild Papers; Richard W. G. Welling, MS. Diary, November 16, 28, 1896, NYPL.

13. Arndt, "Quarter Century of Citizens' Union," p. 6.

14. Albert Shaw to Dr. W. H. Tolman, March 12, 1897, Mayoralty Campaign, 1897, Papers of Seth Low, Columbia University.

15. Gilder to Low, August 18, 1897, ibid.

16. "Mass Meeting of Citizens' Union, 14th Assy. Dist. Headquarters, July 22, 1897," ibid.

17. "Battle Song of the Citizens' Union," ibid.

18. Brooklyn Eagle, October 10 [1897], Scrapbook V, Low Papers; Low to R. Fulton Cutting, Joseph Larocque, Carl Schurz, June 7, 1897, Mayoralty Campaign, 1897, ibid.; Benjamin R. C. Low, Seth Low (New York, 1925), pp. 20-48; Harold C. Syrett, The City of Brooklyn, 1865-1898; A Political History (New York, 1944), pp. 106-137; Citizens' Union, Seth Low for Mayor: His Life and Public Record (New York, n.d.), p. 2; Steven C. Swett, "The Test of a Reformer, A Study of Seth Low, New York City Mayor, 1902-1903," New York Historical Society Quarterly, XLIV (January, 1960), 5-9; "Seth Low," Dictionary of American Biography, ed. Dumas Malone (New York, 1933), XI, 449-50.

19. Shaw to Low, June 2, 1897, Jacob H. Schiff to Low, May 25, 1897, Richard W. Gilder to Low, May 27, 1897, Charles Stewart Smith to Low, July 8, 1897, Mayoralty Campaign, 1897, Low Papers.

20. Low to Cutting, Larocque, Schurz, June 7, 1897, Low to Smith, July 10, 1897, Smith to Low, July 29, 1987, Pamphlet inviting Brooklyn citizens to conference to bring about nomination of Seth Low to be held at Criterion Theatre June 30, 1897, ibid.; James B. Reynolds to Fairchild, August 17, 1897, Box 7, Fairchild Papers; Citizens' Union, Campaign Book, 1897, p. 113;

New York World, July 1, August 19, 21, 27, 31, September 2, 1897.

21. September 9, 1897, Mayoralty Campaign, 1897, Low Papers.

22. Low to H. H. Beebe, September 7, 1897, ibid., Brooklyn Eagle, October 10 [1897], Scrapbook V, ibid.; New York World, September 28, 29, October 1, November 3, 1897; Arndt, "Quarter Century of Citizens' Union," p. 10.

23. John T. Hettrick, "Reminiscences," CUOHC, p. 47.

24. H[enry] L[oomis] N[elson], "Tammany on Trial--The Mazet Investigation," Harper's Weekly, XLIII (September 9, 1899), 881.

25. Ibid., p. 882; Franklin Matthews, "The Cost of Tammany Hall in Flesh and Blood," Harper's Weekly, XLIV (October 13, 1900), 957-60.

26. "The Old, Old Story" (editorial), ibid., March 24, 1900, p. 262.

27. Arndt, "Quarter Century of Citizens' Union," pp. 10-13; New York World, February 23, 1898; Minutes of the City Committee of the Citizens' Union, September 24, 1900, Box K-1, CU Coll. (hereafter cited City Committee Minutes).

28. City Committee Minutes, September 24, 1900. Among the signers of this manifesto were Charles S. Fairchild, Richard W. Gilder, Abram H. Hewitt, William B. Hornblower, Joseph Larocque, Gustav H. Schwab, Simon Sterne, E. R. A. Seligman, J. Kennedy Tod, Everett P. Wheeler, William J. Schieffelin, Jacob A. Riis, Alexander E. Orr, George F. Peabody, and John Claflin; New York World, November 13, 1900.

29. City Club, Report of the Council, December 5, 1900, pp. 7-8, City Club Collection, NYHSL; New York World, November 9, December 11, 29, 1900, February 9, April 10, 26, May 9, 29, 1901; Welling, Diary, April 26, 1901; Minutes of the Executive Committee of Twelve of the Committee of One Hundred of the Citizens' Union, May 14, 1901, Campaigns, Box H-1, CU Coll. (hereafter cited Exec. Comm. of Twelve Minutes); General Committee Meeting of German-American

Reform Union, May 24, 1901, Platform of the Citizens' Union, 1901, Citizens' Union, Shepard-Low Campaign, 1901, Scrapbook I, Material Collected by Everett P. Wheeler, NYPL.

30. Minutes of First Conference Meeting Called by Committee of One Hundred of the Citizens' Union, May 28, 1901, Campaigns, Box H-1, CU Coll. (hereafter cited First Conference Minutes).

31. Minutes of the Committee of One Hundred of the Citizens' Union, May 4, September 4, 23, 1901, Exec. Comm. of Twelve Minutes, August 29, September 4, 1901, Minutes of Full Conference of Anti-Tammany Organizations, September 18, 20, 1901, ibid.; New York World, August 29, September 18, 19, 1901.

32. Citizens' Union, Campaign Book, 1902, p. 9, 1903, p. 10; Allan Nevins and John A. Krout (eds.), The Greater City New York, 1898-1948 (New York, 1948), p. 68; Arndt, "Quarter Century of Citizens' Union," p. 15; Swett, "Test of a Reformer," p. 34; D. S. Alexander, Four Famous New Yorkers (New York, 1923), pp. 398-99; New York World, August 29, September 10, 23, October 6, 1903.

33. New York World, August 16, 29, September 10, October 6, November 4, 1903; Arndt, "Quarter Century of Citizens' Union," p. 15.

34. "The Municipal Campaign" (editorial), Harper's Weekly, XLV (October 19, 1901), 1045.

35. Carl Schurz, "Bossism in New York," ibid., November 13, 1897, p. 1119.

36. "Our Kings" (editorial), ibid., XXXVII (September 30, 1893), 926; "Municipal Campaign," ibid., p. 1045; "How Croker Rules" (editorial), ibid., November 4, 1893, p. 1047; "Tammany Hall" (editorial), ibid., February 11, 1893, p. 122; Carl Schurz, "The Task of the Citizens' Union," ibid., XLI (December 11, 1897), 1215.

37. "A Practical Step in Reform," ibid., XXXVII (November 18, 1893), 1102.

38. Richard W. G. Welling, Purpose of Good Government Clubs; Good Government Club "D" (Election brochure), October, 1894, Political Papers, Box 23, Welling Papers.

39. Bernheim, "Committee of Seventy," p. 9; Knerr, "Mayoral Administration of Strong," pp. 2-30; "The Committee of Seventy" (editorial), Harper's Weekly, XXXVIII (September 22, 1894), 891.

40. Arndt, "Quarter Century of Citizens' Union," pp. 3-8.

41. Charles Garrett, The LaGuardia Years, Machine and Reform Politics in New York City (New Brunswick, N.J., 1961), p. 38; Welling, Purpose of Good Government Clubs, October, 1894, Welling Papers; "Good Government Clubs," Harper's Weekly, XXXVII (June 3, 1893), 519; Executive Committee Meeting Minutes of the City Reform Club, May 11, 1892, NYPL; Richard Welling, As the Twig Is Bent (New York, 1924), p. 41; Minutes of Third Conference Meeting called by the Committee of One Hundred of the Citizens' Union, September 9, 1901, Campaigns, Box H-1, CU Coll.

42. "Battle Song of Citizens' Union," Low Papers; Harper's Weekly, XXXIX (January 19, 1895), 50; "Shepard's Dangerous Good Character" (editorial), ibid., XLV (October 26, 1901), 1067.

43. "The City Club" (editorial), XXXVIII (May 6, 1893), 414.

44. "Shepard's Good Character," Harper's Weekly, p. 1067.

45. Welling, Diary, November 24, 1896; Welling, Twig, p. 57.

46. Welling, "City Reform Club"; "The Good Government Clubs and the Campaign of 1895" (editorial), Harper's Weekly, XXXIX (November 2, 1895), 1032; "The Committee of Fifty," ibid., October 19, 1895, p. 984.

47. Good Government Campaign Headquarters, Document No. 2, October 16, 1895, Political Papers, Welling Papers.

48. Welling, Diary, June 19, 1897; Low to Gilder, September 8, 1897, Mayoralty Papers, 1897, Low Papers.

49. Schurz, "Task of Citizens' Union," p. 1215.

50. Quoted in Werner, Tammany Hall, pp. 328-329.

51. William L. Riordan, Plunkitt of Tammany Hall (New York, 1963), p. 17.

52. Knerr, "Mayoral Administration of Strong," pp. 273-275; "Duties of Public Spirit" (editorial), Harper's Weekly, XXXVIII (December 8, 1894), 1154.

53. Arndt, "Quarter Century of Citizens' Union," p. 13.

54. Knerr, "Mayoral Administration of Strong," pp. 271-275; Schieffelin, "Reminiscences," pp. 18-19; Welling, Diary, November 2, 1897; Garrett, LaGuardia Years, pp. 39-41; Arndt, "Quarter Century of Citizens' Union," p. 13.

55. Riordan, Plunkitt, p. 19.

56. Knerr, "Mayoral Administration of Strong," pp. 271-273; Swett, "Test of a Reformer," pp. 35-36, 40; "Why Municipal Reform Succeeds in Chicago and Fails in New York," Independent, LVI (April 14, 1904), 830.

57. "Battle Song of Citizens' Union," Low Papers; First Conference Minutes, May 28, 1901.

58. Welling, Diary, November 2, 1897; Knerr, "Mayoral Administration of Strong," pp. 253-258; 271-273; Garrett, LaGuardia Years, p. 40.

59. Municipal Reform," Independent, pp. 830-831.

60. Knerr, "Mayoral Administration of Strong," pp. 56-116, 139-196, 261; Citizens' Union, Campaign Book, 1897, pp. 3-8, 47-53; Swett, "Test of a Reformer," pp. 21-26, 35; "Municipal Reform," Independent, p. 830; Nevins and Krout, Greater City, p. 69.

61. Swett, "Test of a Reformer," pp. 35-36, 40; "Tammany Hall," Harper's Weekly, p. 122; Citizens' Union, Campaign Book, 1897, pp. 7-8, 113-15; Clean Streets, Citizens' Union Pamphlet No. 2, 1897, pp. 3-14, Mayoralty Campaign, 1897, Low Papers.

62. Riordan, Plunkitt, pp. 25-26, 28.

63. Welling, Twig, p. ix.

Chapter 3

THE BATTLE AGAINST THE SLUMS

Housing, poverty, and health were interrelated and old problems in New York. Efforts for improvement in these areas overlapped as a host of reformers--physicians, social workers, ministers, politicians, and businessmen--frequently occupied common ground. The slum environment was to these reformers what Tammany Hall was to the good government crusaders--a major source of the ills plaguing the city. If evil men in government needed to be replaced by moral men, slum dwellers needed to be morally transformed. If government and politics needed to be purified, so did impoverished neighborhoods. In practice, reformers worked to improve living conditions as well as change lives.

For much of the nineteenth century New York's business and professional leaders, supported by a host of humanitarians, provided the major thrust behind slum reform. Many were convinced that housing improvement was "the basis of other reforms": the moral, social, and physical uplift of tenement residents.[1] Better housing rarely was considered an end in itself; it was a precondition for rehabilitating lives.

For doctor-reformers, who until the mid-eighties believed disease was caused by dirt and generally poor environmental conditions, cleaner and more healthful tenements meant less sickness.[2] In "A Brief View of the Sanitary Condition of the City," a pioneering and penetrating study in 1842 of New York's burgeoning slums, Dr. John H. Griscom, a physician and perceptive city health inspector, wrote that overcrowded living conditions and the absence of privacy were "prolific sources of moral degradation and physical suffering."[3] The Association for Improving the Condition of the Poor,[4] created in 1843 by a number of the city's leading merchants and businessmen, reported a decade later that "the laboring classes are made filthy, reckless, and vicious by the force of circumstances over which, in most cases, they have no control."[5] Yet this relief agency hoped not merely "to alleviate wretchedness, but to reform character." In an era when moral crusading was a standard reform tactic, the AICP conceived of its primary task as moral evangelism among the city's less fortunate.[6]

Toward the end of the century Jacob Riis, a journalist whose writings exposed and publicized housing and slum problems, preached that tenement lives needed rescuing. Along with his friends in Manhattan's settlement houses, he sought such neighborhood improvements as better housing, schools, playgrounds, parks, youth clubs, and health facilities hoping that these environmental changes would help eliminate the saloons, poolrooms, gangs, and other culturally depriving influences of the ghetto.[7]

A number of reformers, especially the doctors, urged the city and state to take some responsibility for the living conditions of the poor. In 1842 Dr. Griscom argued that "a due regard for the health of the citizens and residents" justified municipal regulation of housing conditions.[8] The AICP suggested that such matters of life and health were community responsibilities. Its merciless exposure of housing and sanitary conditions slowly aroused the legislature which in 1856 appointed a committee of its own members to investigate the tenements. For the first time the state became involved in housing reform. It took the lawmakers only seven days to conclude that legislation was necessary to remedy evils "startling and painful to behold." In March, 1857, the committee presented New York's first legislative report on housing. The report called for the creation of a board of commissioners to regulate the structural and sanitary conditions of tenements and to keep statistics and records of the tenements and their occupants. This far-sighted scheme for a permanent body that would involve the state continuously in slum problems was rejected by the legislature; it had to await a more congenial era.[9]

The brutal and destructive riot against a Civil War draft law in the summer of 1863 provided the immediate impetus for an organized private attack on the slums. Fearful of losing jobs and social status to northernbound emancipated blacks, New York's white immigrants, especially the Irish, openly resisted what they believed to be a discriminatory draft, attacked blacks and Republicans, and destroyed life and property in the city with abandon. That these crude mobs, "furious as demons," were inhabitants of the city's slums was not unnoticed by Gotham's middle-class citizens.[10] The dire warnings of the legislative committee of 1856 that the evils existing in the slums would poison and disrupt not just the foreign ghettos but the whole community and threaten American values seemed prophecy fulfilled. "To walk the streets as we walked them in those hours of conflagration and riot was like witnessing the day of judgment, with every wicked thing revealed, every sin and sorrow blazingly glared upon, every hidden abomination laid before hell's expectant fire," wrote a leading journalist of the day.[11]

The following year the Council of Hygiene and Public Health was created by the Citizens' Association. Organized in 1863 by prominent merchants, lawyers, and men of wealth such as John Jacob Astor, Jr., August Belmont, Hamilton Fish, and Peter Cooper, the Association was New York's first permanent nonpartisan civic group and the predecessor of many similar organizations that battled Tammany during the last three decades of the nineteenth century.[12] The Council, with many prominent physicians as members, was a model for uniting behind a reform cause professional expertise and the prestige and money of the city's businessmen-reformers. This was a pattern not to become prevalent in reform until the progressive years early in the next century. The doctors' primary target was the city health inspector's office. The inspectors, as well as their army of sanitary wardens, generally were political appointees and ignorant of medical or sanitary practices. Besides, the inspector's office had many responsibilities unrelated to health and sanitation.[13]

When the Council of Hygiene's petition for a new city health department was rejected in 1864, the Council decided on a thorough and factual sanitary investigation of Manhattan. Under the direction of Dr. Stephen Smith, a distinguished surgeon, author, and public health enthusiast who later became the first president of the American Public Health Association, the Council divided the island into twenty-nine districts, placed in each an experienced physician as sanitary inspector, and for nine months conducted "the most thorough, complete, and scientific sanitary inspection ever made of any city." Working from house to house, these dedicated doctor-reformers accumulated a mass of data. Tenement houses were counted, inspected, and described, many even photographed, and their inhabitants interviewed; the sickness and mortality rate in each tenement was recorded. Even the type of soil and condition of the streets in each district was described. One of the worst neighborhoods, the fourth ward along the East River in lower Manhattan, was studied with exceptional intensity by Dr. Ezra R. Pulling. "Seldom has a piece of social or sanitary work been better done," Lawrence Veiller, a housing expert, commented about the whole investigation.[14]

"Unless the resources of Sanitary Science and the beneficent operations of widely- administered sanitary regulations are interposed," the doctors reported, "the evils attendant upon overcrowding and the aggregation of vast numbers will be continually augmented as the population increases."[15] Blaming slum conditions for the plight of the poor and recognizing that this class of New Yorkers, the majority of whom were newcomers to America and urban living, possessed no leverage to get better sanitary facilities, larger apartments, better building plans, and more light and air, the doctors placed the responsibility on city hall. "Forethought, inquiry, science, art and good governmental regulations" should characterize responsive government, they pleaded.[16]

The Council, which published the fruits of its labor in a report of 360 pages, provided an example of the new reform approach it urged upon city administrators. These physicians brought to the solving of urban social issues the same penchant for research and reporting that increasingly characterized medical investigation. The Council of Hygiene also recognized that adequate machinery was needed to enforce reform measures and thus again pushed in Albany for a new department of health. The state's lawmakers were less impressed with Dr. Smith's data and arguments for a new department than with the city inspector's pressure for rejection. Only the scare of a cholera epidemic in 1865 panicked the legislators into passing a law that created the Metropolitan Board of Health.[17]

The Board of Health was a permanent, continuously functioning state body with jurisdiction in New York City; that is Manhattan and the Bronx. Of the nine commissioners comprising the board, four were appointed by the governor, three of whom had to be physicians. The health officer of the Port of New York and four metropolitan police commissioners completed the membership. By law the board was compelled to appoint a physician as sanitary superintendent and likewise ten of fifteen sanitary inspectors appointed had to be doctors. Thus, the lawmakers recognized the need for professional intelligence in public administration. However, the doctors' influence was minimized because the board was constantly thwarted in its efforts by New York's landlords and builders who were influential with Tammany Hall.[18]

The year following the creation of the Board of Health, the state legislature, after an inquiry of tenements in Brooklyn and Manhattan which corroborated the findings of the Council of Hygiene, passed New York's first tenement house law. The new law was essentially a rule book, but the required standards for tenements were low and many of the provisions vaguely worded. Most of the rules relating to sanitary conditions, cellars, and ventilation provided exemption clauses or could be significantly modified by the Board of Health, the agency given the task of enforcing the new tenement law. Houses containing up to three families were exempt. Yet slum living was no respecter of numbers.[19] If not in practice at least in principle the first tenement house law "represented the acceptance . . . of the community's right to limit the freedom of the tenement landlord and builder."[20] This act and the previous one creating the Board of Health established the pattern, modified only in detail, for government regulation of housing for the remainder of the nineteenth century.

In 1878 the State Charities Aid Association capitalized on a revival of public interest in housing reform. This renewed interest was sparked by the lectures and model tenement building of Alfred T. White, a wealthy Brooklyn businessman and philanthropist, and a contest for the best architectural design for a tenement house sponsored by several businessmen and Henry C. Meyer, a manufacturer of plumbing supplies and proprietor of the trade journal Plumber and Sanitary Engineer. The Charities Aid Association sponsored several housing conferences, appointed a special committee to study housing needs, and recommended that the tenement laws be more rigidly enforced and amended.

In February of the next year a tenement house Sunday was proclaimed and the city's clergymen, with a fervor equal to that of the ancient Hebrew prophets, preached against the evils of the slums and pleaded for reforms. To arouse even further the community's interest, the next week the Charities Aid Association sponsored a public rally at Cooper Union, that mugwump temple, which was presided over by the mayor and attended by New York's leading citizens. A committee of nine was appointed to shape a reform program. Upon its recommendations, the Improvement Dwelling Association, with Bayard Cutting, a member of the committee, as president, was formed to build model tenements, and the New York Sanitary Reform Society was organized as a permanent private association to work to improve housing conditions. The committee also prepared and had introduced in the legislature a bill to amend the tenement house law of 1867. In spite of vigorous opposition from the state's landlords, the bill became law in 1879.

The new tenement house law did little but alter a few of the rules in the original housing bill. To improve enforcement the act provided funds for thirty sanitary policemen to be detailed from the police to the health department. Unfortunately a new provision that every sleeping room have at least one window admitting light and air directly from the street or yard also permitted the Board of Health to approve other ventilation schemes. This paved the way for the widespread adoption of the dumbbell tenement designed by James E. Ware and awarded first prize in Meyer's tenement house contest. Ware's building was a front and rear tenement connected by a narrow hall that gave the entire building the shape of a weight-lifter's dumbbell. Each floor of the five- or six-story building contained fourteen rooms divided into four apartments. Most of the rooms received light and air only from a narrow shaft on each side of the building. These air shafts appeared to meet the requirement of the amended tenement law. Unwittingly, therefore, the reform act of 1879 encouraged the speedy adoption by investors of the dumbbell plan. To the landlord's financial benefit, Ware's house fulfilled the letter of the law but certainly killed its spirit. In little better than twenty years Veiller referred to the dumbbell house as the curse of the city.[21]

The amended tenement law quieted reformers for a while, but neither it nor the work of the Board of Health significantly altered the living conditions of the poor. Five years later Professor Felix Adler, founder of the Ethical Culture Society, delivered a series of lectures denouncing New York's slum housing. Again the press and community were aroused, and this time the legislature appointed a tenement house commission. Adler and Charles F. Wingate, a civil engineer, former editor of Plumber and Sanitary Engineer, and authority on housing problems, were members of this commission which began to investigate tenement conditions in June, 1884. The investigators published a report of 235 pages including twenty recommendations for rather superficial changes in the housing law. The law of 1887 did not affect the basic pattern of tenement living or the method of regulating slum conditions. It provided for a standing tenement house commission, composed of the mayor and the heads of the Departments of Health, Public Works, and Street Cleaning, to meet yearly to discuss tenement problems. The commission was ineffectual, but the idea was potentially important.[22]

In 1894, the year the Committee of Seventy organized to elect William Strong as mayor, the New York Press launched a series of articles describing the still deplorable tenement conditions. Edward Marshall, its Sunday editor, successfully led the fight for a bill authorizing the governor to appoint a state commission to examine the slums. The third of such commissions appointed since mid-century, it included as chairman Richard Watson Gilder, well-known magazine editor, man of letters, and supporter of various mugwump causes; William D'H. Washington, a civil engineer; George B. Post, an architect; Roger Foster and John P. Schuchman, lawyers; Solomon Moses of the United Hebrew Charities; and Dr. Cyrus Edson, sanitary superintendent of the Board of Health. During the course of its work the Gilder committee leaned heavily on the advice of Jacob Riis.

The Gilder committee began to work in May and the following January published an extraordinary report of over six hundred pages that dealt with the whole complex of social issues plaguing the tenement districts--lack of neighborhood recreational and sanitary facilities such as parks, baths and toilets, immigration, population density, rents, housing construction, model tenements, death rates, prostitution, sweatshops, rapid transit services, and fire hazards. The report included maps and photographs of tenement areas and illustrations of various tenement house plans. "The committee's careful and comprehensive analysis," writes Lubove, "resulted in a valuable civic education for New York."[23]

To the slum dwellers' disadvantage, the Gilder committee earned higher marks for its report than for its achievements. Its recommendations, many of which were transcribed into the tenement house law of 1895, barely scratched the surface of slum problems and did nothing to alter or even retard tenement development. Responsibility for housing regulation was divided among the health, buildings, and fire departments, and the committee ignored this unsatisfactory bureaucratic arrangement. Its major concession to administrative change was to perfect the law granting to the Board of Health power, under certain circumstances, to condemn and have destroyed houses unfit for living because of

> age, defects in drainage, plumbing, infection with contagious disease, or ventilation, or because of the existence of a nuisance on the premises, which is likely to cause sickness among its occupants or among the occupants of other property in the city of New York, or because it stops ventilation in other buildings, or otherwise makes or conduces to make other buildings adjacent to the same unfit for human habitation, or dangerous or injurious to health; or because it prevents proper measures from being carried into effect for remedying any nuisance injurious to health or other sanitary evils in respect of such other buildings. . . .[24]

Only if such conditions could not be remedied could the board, through precise legal procedures and provision for compensation to the owner, have a house destroyed. Other minor administrative changes included an increase in the number of sanitary inspectors and police, the closing of a loophole in the law requiring tenement owners to register with the Board of Health, and the abolition of the ineffective tenement house commission created by the act of 1887.

Recommended structural and sanitary changes enacted into law included the prohibition of certain dangerous trades in tenements not fireproof. In addition, tenements were not to be used as lodging houses, stables or storage places for rags, every water closet was to have one window to the outer air and a watertight floor, each room in new houses was to have one window opening to the outer air, and all cellars had to be two feet above the ground. To reduce overcrowding the Board of Health was to enforce the law that each adult have at least 400, and each child under twelve, 200, cubic feet of air, and no new tenement or enlarged old one could occupy more than 75 percent of a lot. The committee also recommended and ultimately saw adopted plans for small parks in the lower East Side, outdoor playgrounds as part of new school buildings, and recreation piers along the city's rivers. Also suggested but largely ignored were pleas for more rapid transit facilities, public baths, lavatories and drinking fountains, and better paved streets in slum neighborhoods.[25]

As a package of restrictive legislation, the committee's recommendations and the law of 1895 changed little in a housing code essentially written during the preceding few decades. Private baths and toilets in each apartment, not public facilities or windows in private water closets, were sorely needed. Restricting the lot coverage for tenements was a nice gesture, but the standard lot, twenty-five by one hundred feet, was much too small. Condemning exceptionally poor houses and enforcing rules to limit the inhabitants per room was fine from the standpoint of health and sanitation, but where were thousands of immigrants to live and how would countless others, prohibited from renting sleeping space, pay the rent?[26]

Because no one knew what a satisfactory tenement might be, reformers as well as lawmakers were stabbing in the dark. It is little wonder that their handiwork failed to prevent the expansion of the evil tenement system. Most of the standards incorporated in the succession of tenement laws barely touched upon the inadequacies of the tenement as a house fit for human habitation. Too many tenants were crammed into tiny apartments. Light and air were scarce commodities. To provide decent housing for large numbers of people on the small city lot was impossible. Either the tenants' welfare or a portion of the landlords' profits had to be sacrificed. Too few housing reformers, however, many of whom were businessmen committed to private enterprise, thought to condemn the small city lot.[27] After several decades of agitation, housing studies, sanitary surveys, and legislation, the tenement house law of 1895 still amounted to only a few rule changes in an old handbook.

Even the Gilder committee, which so thoroughly studied slum conditions, suggested no new and perhaps more effective methods for dealing with the tenement blight. Like their counterparts in the good government crusades, housing reformers in the nineteenth century did not envision an enlarged city and state bureaucracy with continuous responsibilities for urban ills. Consequently they rarely asked whether the existing administrative methods were adequate to regulate tenements and enforce the laws. The under-staffed Board of Health had multiple responsibilities, including street cleaning and the preparation of a city health code. Yet policing tenements alone would have required a full-time staff. The board possessed no power to force compliance with the law but had to rely on the legal system, a slow and uncertain process.[28] Not until the turn of the century did a professional housing reformer, Lawrence Veiller, come forward with a workable plan for a public agency to deal solely with the mammoth slum housing problem.

Furthermore, the reformers' own efforts were discontinuous. The Citizens' Association was a temporary grouping of good men to do battle for cleaner government. The AICP and State Charities Aid Association were engaged in a number of relief activities, and their housing reform efforts were limited. Demands for legislative investigations were sporadic. Occasional public lectures, sermons, and rallies and newspaper exposes of poor housing were worthy endeavors in civic education but certainly no substitute for continuous study of and involvement in tenement problems. Housing reformers in the nineteenth century acted very much like Plunkitt's "mornin' glories." Occasionally they engaged in a cooperative act of goodness by attacking the slum evil with a new law, but then returned to the sides sure that their righteous act would bring about cleaner neighborhoods. However, by 1900 approximately two-thirds of New York's inhabitants still lived in tenements that were overcrowded and filthy. Just as New York's respectables, after an occasional victory at the polls, witnessed the inevitable return to power of Tammany, slum reformers saw their feeble efforts at improvement overshadowed by tenements that continued "to pack Manhattan's east side and west side . . . [and extend] to new sections in Brooklyn and the Bronx."[29]

Yet the Gilder committee and its work at the end of the century represented a transition in housing and sanitary reform. Its factual portrayal of New York's seamier side was a triumph in civic education and reflected the realism and scientific spirit that characterized the emerging public health movement and presaged later progressive activity. While Riis, as an influential advisor, and Gilder, as leader of the commission, represented the nonprofessional, nontechnical lay approach to reform, the doctor, architect, engineer, and two lawyers on the commission represented specialties concerned with certain technical aspects of slum improvement. It was only a matter of time before the professionals would take charge and redirect the entire slum reform campaign.

NOTES

1. Lilian Brandt, Growth and Development of AICP and COS (New York, 1942), p. 48.

2. George Rosen, A History of Public Health (New York, 1958), p. 239.

3. Lawrence Veiller, "Tenement House Reform in New York City, 1834-1900," in The Tenement House Problem, eds. Robert W. deForest and Lawrence Veiller (2 vols.; New York, 1903), I, 71-75.

4. Hereafter referred to as AICP.

5. Quoted in Veiller, "Tenement House Reform," p. 81; Roy Lubove, The Progressives and the Slums: Tenement House Reform in New York City 1890-1917 (Pittsburgh, 1962), pp. 4-7.

6. Brandt, Growth and Development, pp. 11-21.

7. Harold S. Buttenheim, "Jacob A. Riis--Valiant Foe of the Slum," May 3, 1949, Box 7, Papers of Jacob A. Riis, Library of Congress; Lubove, Progressives and Slums, pp. 49-80; Jacob A. Riis, The Making of an American, ed. Roy Lubove (New York, 1966), pp. xii-xxvi

8. Veiller, "Tenement House Reform," p. 75.

9. Ibid., pp. 87-91.

10. Lubove, Progressives and Slums, pp. 11-12; A. Hunter Dupree and Leslie H. Fishel, Jr. (eds.), "An Eyewitness Account of the New York Draft Riots, July, 1863," Mississippi Valley Historical Review, XLVII (December, 1960), 472-479.

11. Quoted in Veiller, "Tenement House Reform," p. 92.

12. Garrett, LaGuardia Years, p. 21; Rosen, History of Public Health, p. 244.

13. Lubove, Progressives and Slums, pp. 11-16.

14. Veiller, "Tenement House Reform," pp. 93-94.

15. Quoted in Lubove, Progressives and Slums, p. 21.

16. Quoted in ibid., p. 22.

17. Ibid., pp. 22-23; Veiller, "Tenement House Reform," p. 94.

18. Rosen, History of Public Health, pp. 246-247; Lawrence Veiller, "A History of Tenement House Legislation in New York, 1852-1900," Appendix VI of The Tenement House Problem, II, 205; Lubove, Progressives and Slums, p. 27.

19. Lubove, Progressives and Slums, pp. 25-28; Veiller, "Tenement House Reform," pp. 94-97; Veiller, "History of Tenement House Legislation," p. 204.

20. Lubove, Progressives and Slums, p. 27.

21. Ibid., pp. 28-33; Veiller, "Tenement House Reform," pp. 97-102.

22. Lubove, Progressives and Slums, pp. 29, 43; Veiller, "Tenement House Reform," pp. 103-104.

23. Lubove, Progressives and Slums, pp. 88-90; Veiller, "Tenement House Reform," pp. 105-106; Veiller, "History of Tenement House Legislation," p. 335.

24. Veiller, "History of Tenement House Legislation," p. 317.

25. Ibid., pp. 208ff; Veiller, "Tenement House Reform," pp. 106-107.

26. Lubove, Progressives and Slums, pp. 91-93; 95-99, 114.

27. Ibid., pp. 31, 32; Veiller, "Tenement House Reform," pp. 97-102.

28. Veiller, "Tenement House Reform, pp. 94-97; Veiller, "History of Tenement House Legislation," p. 204; Lubove, Progressives and Slums, pp. 26-28.

29. Edward T. Devine, When Social Work Was Young (New York, 1939), p. 69.

Chapter 4

THE PROMOTION AND REGULATION OF UTILITIES

Few economic functions so continuously influenced the living and working habits of all New Yorkers as did utility services. Because gas, electric, telephone, and transportation facilities were privately owned, concerned citizens, ranging from leading businessmen to university-trained professionals, insisted on the public's right to limit this private "power to tax the common necessities of life."[1] But in the nineteenth century few New Yorkers understood the public's right to mean granting continuous and enlarged managerial powers over utilities to an expanded city or state government. Instead concerned individuals put faith in competition, restrictive franchise provisions, and special legislative investigations and laws. Such regulatory methods were advocated in the hope that singly or together they would ensure good utility services at low rates and protect the city's financial stake and sovereignty in the streets.

Neither competition nor franchise stipulations proved adequate instruments of control. Nor was legislation very effective. Yet the many state investigations conducted, commissions created, and laws passed represent the more significant nineteenth-century contributions to utility improvement. They provided a pattern for action easily extended by a later generation of reformers. These efforts over almost half a century were not parts of an evolving policy but were discrete responses to immediate problems.[2]

Rapid transportation was New York City's most persistent public utility problem. A number of reformers, primarily businessmen concerned about keeping New York the commercial and financial center of the nation, led the search for a more efficient rapid transit service. Because no municipal agency alone was responsible for the city's utility needs, these men of means channeled their influence through the Chamber of Commerce and other business organizations. In the process they stamped their values upon the quest for more transportation.

Businessmen divided on two important questions. Should elevated lines or subways be promoted as the backbone of a rapid transit system? Should transportation be municipally owned and operated, municipally owned and privately operated, or privately owned and operated? In 1864 Abiel A. Low, William E. Dodge, Jr., and John Jacob Astor, Jr., among other business leaders, organized a company to build a subway but failed to get a franchise from the state legislature. Other businessmen, most notably Abram S. Hewitt, ironworks master and future mayor of New York, shared the fervor for a subway but wanted it built and operated by the city. This suggestion likewise failed.[3]

A state senate committee studying New York City's transit needs in 1864 also suggested that a subway was the "only speedy remedy for the present and prospective wants of the city of New York, in the matter of the safe, rapid and cheap transportation of persons and property."[4] But it also asked the legislature to help finance a half-mile experimental elevated cable railroad that had been proposed by Charles T. Harvey, an engineer. If successful, the experimental section was to be the first link in an elevated road running along Ninth Avenue from the Battery to the Harlem River. The legislature did nothing about the subway but approved Harvey's scheme and unknowingly closed the door to underground rapid transit for the remainder of the century. Harvey's experiment with a cable-propelled railroad proved impractical beyond the initial stage, and his company, bankrupt by 1871, was acquired by the New York Elevated Railroad Company which successfully substituted steam as the motive power.[5]

In 1874 Mayor William H. Wickham, concerned about "the growth and general prosperity of the city," and the Board of Aldermen joined the mushrooming debate among the city's businessmen over the nature and proper ownership of new rapid transportation.[6] The next year the state legislature responded to pleas for help and passed a rapid transit act which made possible New York City's first rapid transit commission. Actually, the act of 1875 placed the initiative for new facilities with New York's citizens. Only after fifty or more householders and taxpayers petitioned for steam railways could the mayor appoint five commissioners to determine the need for the roads and to adopt routes "over, under, through or across the streets, avenues, places, or lands." The board of commissioners was granted power to fix maximum rates, to organize new construction companies, and to allow existing elevated companies to build connections with other steam railways and ferries. After the board approved a route and company to build and operate a transit line it ceased to exist.[7]

The first commission appointed under the terms of the act of 1875 decided on an elevated system and created the Manhattan Railway Company to build the new road. This decision served to fasten on the metropolis for the next three decades the noisy, ugly, and obstructionist el. During these years other commissions were chosen and elevated railroad companies organized. However, by the turn of the century all of these companies had merged into the Manhattan Railway Company making this private creature of a public rapid transit commission sole owner of all elevated lines in old New York.[8]

To the consternation of New Yorkers, the clogging of several Manhattan streets with els failed to meet the need. "The existing railroads have practically reached the limit of their capacity," Mayor Hewitt lectured the state legislature in 1888. "Besides . . . they are not in reality satisfactory to any class of the community. They do not in effect provide rapid transit, which is the chief requirement of that portion of our population who night and morning pass from one end of the island to the other."[9] Hewitt, who a decade before had advocated a municipally owned subway, now suggested combining the credit of the city with private construction and operation. Although legislation embodying this idea received support from the Real Estate Board, the Chamber of Commerce, and even the County Democracy and a few Tammany leaders, it died in the legislature. Out of office the following year, Hewitt continued to support rapid transit improvement through the Chamber of Commerce, and in less than a decade he was to see his plan come to fruition.

In 1891 the legislature passed a law providing for New York's first permanent rapid transit commission. It was given power to grant additional franchises to existing railroad companies and to determine if new rapid transit roads were necessary. In the event a new road was authorized the commissioners were to select routes, specify the types of construction and operation, obtain the consent of the local authorities and affected propertyholders, and then for a "period of time as the commission should deem advisable," sell such rights to the highest bidder. Significantly, municipal construction and ownership of rapid transit was not contemplated.[10]

For the commission and other private organizations working to improve rapid transportation, the proper mix between elevated roads and subways was almost as perplexing and divisive a question as municipal versus private ownership. When the rapid transit commission failed to entice private entrepreneurs to build a proposed subway, it decided to pursue as an alternative the expansion of Jay Gould's elevated system, only to be warned by one of its members, John Starin, that if the board gave away too many privileges it would permanently seal the doom of rapid transit. "You cannot think of having two rapid transits in the city of New York," he argued.[11]

Conflicting advice bombarded the commissioners. A majority of the Real Estate Exchange, labor unions, and a number of civic groups opposed extending the elevated system and argued for a municipally owned subway. The influential Board of Trade and Transportation was divided over the issue but at the urging of two of its members, Francis B. Thurber and

Simon Sterne, came out against a municipal subway. Tammany Mayor Thomas F. Gilroy and other city officials likewise opposed any move to give New York the power to build its own transportation facilities. They claimed that if the city built a road the debt incurred would exceed the constitutional limit of 10 percent of the assessed valuation of New York's taxable real estate.

Precluded by law from building a subway with city funds, unable to convince a private builder to undertake the task, and unwilling to seek the right of city ownership, the rapid transit board in 1893 reopened negotiations with the Manhattan Elevated Company. When the company refused to accept the Board's suggested rate of compensation to the city from net receipts, John Starin alone voted against reducing the rate of compensation. Because of the unanimous vote clause in the act creating the commission, Starin's vote checked the immediate expansion of the Manhattan system, left the door open for underground transportation, and hastened the demise of the first permanent commission.[12]

The debate over the nature of New York's transportation facilities grew more intense as 1893 came to a close. Richard T. Wilson, a banker acting for a syndicate of capitalists, publicly announced he would build a subway for forty-five million dollars if the city would lend its credit to the amount of thirty million. Because the city was prohibited by the state constitution from lending money or credit to individuals or corporations, the rapid transit commission ignored the proposal. But the idea of using the city's credit to aid a private enterprise served as a catalyst for action by the businessmen-reformers of the Chamber of Commerce. On the first of February, 1894, the Chamber appointed a committee to examine the transit question and recommend to the Chamber a course of action. Two weeks later the committee recommended a subway but without explanation rejected municipal construction. Instead it urged the Chamber to recommend the city lend its credit to a private corporation along the lines proposed by Wilson.[13]

In an extraordinary rebuttal to this report, Abram Hewitt raised the question of the city owning the road outright.

> When the State of New York was confronted by a greater problem than this, the greatest that has ever presented itself to a civilized community on this side of the Atlantic--the construction of the Erie Canal . . . the state took it in hand; became the owner, borrowed the money, paid for the work and, perhaps, unfortunately, retained the administration of the property after it was constructed.
>
> The city of New York has pursued the same system. The Croton Aqueduct was constructed by the public, at its own cost, and administered by its own officers. In no instance has the city of New York gone into partnership with a private company.[14]

It was no time for the city to start now, pleaded the old ironmaster; such a precedent would open the way for numerous speculators to seek city funds for all types of private schemes. Moreover, Hewitt reminded his fellow businessmen, a better alternative was open to the city--the one that he had suggested when mayor. The city could get from the legislature permission to construct a subway and borrow the money at a cost not exceeding 3 percent. To prevent possible graft and payroll padding, a private corporation would construct and operate the road and pay not only interest on the bonds but enough rental to establish a sinking fund to retire the city's debt. Needing more time to consider the committee's and Hewitt's plans, the Chamber postponed a decision until the next month's meeting.

At the meeting in March, Hewitt offered his plan as a substitute for the committee's recommendations, and it was adopted after only minor criticism. Even Wilson openly favored the new plan as more feasible than his own. The Chamber's rapid transit committee, now including Hewitt and Charles Stewart Smith, president of the Chamber of Commerce, asked Henry R. Beekman to draft a bill embodying Hewitt's ideas. This bill, modified in some details but with only one major amendment, became law at the close of the legislative session in 1894. Several Chamber of Commerce men lobbied for it before the legislature and governor. Technically a revision of the act of 1891, the new law reorganized the rapid transit commission making it practically a committee of the Chamber of Commerce. In addition to the mayor, the comptroller, and the president of the Chamber of Commerce, who were ex-officio members, the act named to a board of commissioners five men all of whom were members of the Chamber.[15]

The commission retained its power to grant additional franchises to elevated lines and adopt plans for new railroads. In contrast to the original Beekman bill, which left the matter of the city building a subway to the discretion of the board, the final act directed the commission to submit to the voters at the next general election "the question whether such railway or railways shall be constructed by the city and at public expense."[16] If the people of New York voted against using city funds, the rapid transit commission was to sell the franchise to a private corporation as provided in the act of 1891. If the people voted for municipal ownership, the road was to be constructed at municipal expense and remain the absolute property of the city; New York was authorized to issue fifty million dollars worth of bonds to pay for the construction.

The act also stipulated that the same party that constructed the road be required "to lease, equip and operate the railroad for a term of not less than thirty-five nor more than fifty years at an annual rent sufficient to pay the interest upon the bonds to be issued by the city and one per cent in addition thereto."[17] To protect the public interest, the commission, when drawing up the contract, was empowered to determine the details of construction and operation of the road and to specify "conditions as to rates of fare to be charged and the character of service to be furnished."[18]

While these provisions seemingly gave to the board greater latitude in securing a subway for New York, they also limited its powers. If city funds were to be used, the city's debt limit became a necessary consideration before contracts could be let. The duration of any lease granted by the commission was prescribed by law and separate contracts for construction and operation could not be negotiated, even if advantageous to the city. "The motive for this provision," explained Henry C. Wright, secretary of the City Club, over a decade later, "no doubt was the feeling that if a person who was to equip and operate were to construct, it would guarantee a speedy and economic carrying out of the building of the line."[19]

In June, 1894, the new board organized with Alexander E. Orr as president, Henry R. Beekman and Albert B. Boardman as counsel, and William Barclay Parsons as chief engineer. Because Orr was elected president of the Chamber of Commerce that year and thus automatically sat as an ex-officio member of the commission, John H. Inman, a member of the previous commission, filled the vacancy. The commissioners quickly announced that a subway system was the only feasible answer to Manhattan's transit needs and with that decision-making ritual behind them sent Parsons abroad to study European municipal transit systems. After tentatively readopting routes and plans of the commission of 1891, the board submitted the question of municipal ownership to a popular vote during the November election. By more than three to one Gotham's citizens approved city construction of transportation and as if to vindicate the political influence of the Chamber of Commerce and the public role of many of New York's leading businessmen, elected William L. Strong, a businessman and member of the Chamber of Commerce, as mayor.

Early in 1895, the young and energetic Parsons prepared the plans and routes for the proposed subway. To be sure they were technically and financially feasible, the board had the plans examined by a group of experts. After they suggested minor revisions, the board adopted final plans and had them approved by the city council. Legal difficulties, however, resulted in further changes and lost time. In fact, not until March, 1898, was the board able to prepare a completed contract for construction.[20]

In April the rapid transit commission submitted its completed subway contract to John Whalen, the Tammany-appointed corporation counsel, who by law was required to either approve or disapprove its form. For seventeen months Whalen did nothing about the contract, then returned it with suggestions for amendments. His excuse for the delay was the poor financial condition of New York. In 1898, the same year that Tammany again began ruling the city, Greater New York was created under a new city charter. The enlarged city assumed the debts of all the consolidated communities. Because of poor municipal bookkeeping, an accurate estimate of the total debt was impossible to ascertain, but the city comptroller suggested that it probably was over the constitutional limit of 10 percent of the assessed value of real estate. In the administration's view, this doomed the immediate municipal construction of a subway. The city could not legally borrow more money.

The rapid transit board, discouraged by Whalen's procrastination and the financial plight of the new city, again considered having the subway built as a private enterprise. However, it could not get an acceptable enabling measure from the state legislature.[21] The commission then marked time by bombarding the offices of the mayor and corporation counsel with pleas that Whalen approve the subway contract. In the fall of 1899 circumstances aided their cause when the city's borrowing capacity increased. City real estate was revalued at a higher and more realistic level, and a constitutional amendment was adopted which excluded from the city's debt limit about thirty million dollars of county debt. The corporation counsel then approved a revised draft of the contract, five years after New Yorkers first approved a municipally owned subway. The city was to pay for the cost of the road, and the contractor was obligated to build and equip the subway in accord with the commission's specifications and to lease and operate it for a period of fifty years with a renewal option of twenty-five years. Other provisions in the contract dealt with financial and service matters. At the termination of the lease, the city was to buy the equipment and become sole owner of the subway.

Three months after the contract was approved it was awarded to John B. McDonald, a contractor who had come to New York shortly after building a tunnel in Baltimore for the Baltimore and Ohio Railroad. August Belmont supplied McDonald with the necessary funds and organized the Rapid Transit Construction Company which entered into an agreement with McDonald to finance the entire subway. Two years later control of the contract passed to Belmont's newly formed Interborough Rapid Transit Company which in a short time gained control over all transportation facilities in Manhattan. Work on the road formally began in March, 1900, and four years later, October, 1904, the first trains rolled. Two years after the first contract was awarded, the transit commission negotiated with the same company a second contract which extended the Manhattan-Bronx line to Brooklyn. This was opened for service in July, 1905.[22]

Many of the men who dominated the quest for better transportation in New York City also played leading roles in mugwump politics. In the good government campaigns, most reformers, even when squabbling over tactics, agreed that Tammany men needed to be replaced by reputable and honest citizens. Yet in the controversy over a rapid transit system reformers divided even over goals. Because few businessmen supported municipal operation as well as ownership, most of the controversy focused on municipal ownership alone; that is, whether the city should finance the construction of a subway to be built and operated by private interests. Those who opposed even this limited city endeavor generally favored either a wholly private subway or the extension of Jay Gould's Manhattan elevated system.

When the rapid transit board in 1893 vacillated between extending the els and having a subway built, the Real Estate Exchange members gave the commission conflicting advice. While the majority supported the idea of a city-owned subway, a vocal minority dissented and asked that only the privately owned Manhattan elevated lines be extended. Later that same year, however, the Real Estate Exchange unsuccessfully sought legislation amending the act of 1891. The Exchange wanted a new rapid transit commission, liberalized conditions under which private capital could build a subway, the extension of els to the northern limits of the city, and city construction of a subway if private capital refused. Francis B. Thurber, Simon Sterne, and other members of the Board of Trade and Transportation's committee on railway transportation placed that organization in opposition to city ownership in 1893. They feared the risks were too great for the city to construct and possibly operate a subway and the debt too great a burden for New York's citizens. Instead the Board of Trade and Transportation only recommended that existing elevated lines be extended and the companies made to pay the city a fair compensation for new privileges.[23]

In 1894 a number of Chamber of Commerce members wanted the city to lend its credit to a private corporation to build the subway. Private enterprise, "always more or less timid" when great costs are involved, needed to be stimulated, rationalized these businessmen. Because the city could borrow money at lower interest rates, fixed charges would be less and New Yorkers would get cheaper transportation, explained banker Jacob H. Schiff.[24] Abram Hewitt, the leading spokesman for a city-owned subway, opposed this scheme, but even his plan did not envision city operation. His proposal to combine the city's money with private construction and operation was supported by a majority of men in the Board of Trade and Transportation, Real Estate Exchange, City Club, and good government clubs. Mostly private entrepreneurs and independent professionals, these men generally distrusted government, especially one controlled by Tammany. When Hewitt referred to the state-owned Erie Canal to bolster his argument for a city-owned subway, even he confessed that it probably had been a mistake for the state to have "retained the administration of the property after it was constructed."[25] None of these gentlemen thought about restructuring New York's government so as to give the city the means to manage an enterprise as vast as a transportation system.

The utility reform leadership's commitment to this compromise arrangement of public ownership and private construction and operation was never firm. In 1895 the Board of Trade and Transportation cautioned that in spite of strong public sentiment against the private rapid transit companies "there are situations when their interests are in accord with the highest public interest, and public passion or prejudice should not be allowed to blind us and prevent our securing such advantages." By such advantages, the board meant the largest increase in facilities at the lowest cost. Thus in 1899 the Board of Trade and Transportation was willing to consider seriously an amendment sponsored by the rapid transit commission allowing the commission, contrary to the people's wishes, to negotiate for a privately owned road even under a perpetual franchise.[26] The cautious and gentlemanly City Club skirted the municipal ownership question, and the club's council placed that organization only in opposition to the surrender in perpetuity to a private corporation "the most important and valuable franchise in this city." That the club was not opposed to private enterprise undertaking the project was implied in the council's concern "that provisions should be made . . . for repossession [of a subway] by the city at the end of a term of years."[27]

In contrast, the most vociferous opposition to private ownership under any circumstances came from the Social Reform Club. The club organized a committee of fifty to fight the transit amendments giving to the rapid transit board power to negotiate for a private subway and held a debate on the issue at the People's Institute. Felix Adler, William S. Rainsford, James B. Reynolds, and the ubiquitous R. Fulton Cutting, among other well-known humanitarian reformers, signed a Social Reform Club protest against abandoning a municipally owned subway. Various labor unions, including the Knights of Labor, the press, and the ever flexible Tammany Hall, joined the protest that killed the transit board's proposals.[28]

Division within the reform ranks was paralleled by a greater cleavage between the reformers, including the rapid transit commission, and Tammany Hall. When the Chamber of Commerce secured from the legislature the rapid transit act of 1894, which provided for a new commission, these well-to-do gentlemen insulated it from New York City's government. Although the board performed local functions, the members were named in the act. All were leading businessmen, members of the Chamber of Commerce, and active in mugwump politics. Furthermore, the board was self-perpetuating. For these mugwump types, heady notions of home rule and centralized executive responsibility were subordinate to the realities of Tammany politics--or as they put it, "the great republican need of unofficial service"--and the extension of their own influence to city-wide problems.[29] These attitudes certainly were consistent with their own good-government rhetoric. Reputable men needed to purify New York's government by rescuing it from the rule of knaves and partisans--Tammany men. Therefore, keeping subways under private, and in their thinking, purer, management and the rapid transit commission staffed with upright men of breeding were part of the same process.

Tammany Hall knew how to fight back. After Mayor Strong's brief administration ended in 1897, the Tiger again controlled New York behind Mayor Robert Van Wyck. All of the rapid transit commissioners had supported Seth Low, Van Wyck's opponent. This portent of disharmony between business and political members of the commission was crystallized when the new mayor in his first message to the municipal assembly opposed the construction of a subway with city credit and advocated instead extensions of the elevated and surface street railways.[30] Disagreement between the commissioners and the administration became outright hostility in March and April of 1898 when Tammany made a thrust at the commission's very existence. Richard Croker, Tammany's leader, accused the board of doing nothing to expand rapid transit facilities and of being practically an affiliate of the Chamber of Commerce, an organization, he was reported to have said, "led by narrow-gauge men who have only their own selfish interests at stake."[31]

A Tammany senator, Timothy P. Ellsworth, introduced a bill abolishing the rapid transit commission and replacing it with a bipartisan board. This form of city commission was a much used political trick that united the state Republican and city Democratic machines and allowed them to share power in New York. Such a blatant power grab failed to reckon with the spirit of reform politics in the nineties. With the villain so well identified, New York's civic and commercial groups responded by uniting to defeat what the Chamber of Commerce labeled Tammany's move against reform.

At a meeting called by the Chamber a group of leading businessmen and mugwumps signed a petition opposing the Ellsworth bill. A committee of five was organized to lobby against it in Albany. These prestigious businessmen were joined in opposition by the Real Estate Exchange, which saw the bill as injurious to the interests of New York. The City Club was shocked that the bill recognized "politics as an element in the management of a great public work."[32] Not surprisingly, the newly formed Citizens' Union, led by many of the same men who belonged to the Chamber of Commerce, City Club, and other special interest organizations, worked to unite all the opposition to the Ellsworth bill and appointed a committee of fifty to lobby in Albany. Abram Hewitt headed this group. Even labor joined the chorus to save the rapid transit commission. "The history of the degenerate days of the Roman and Grecian republics is being repeated in New York again," cried

Henry White, General Secretary of the United Garment Workers of America. Fortunately for the commissioners, the cause of underground rapid transit, and the egos of New York's "nonpolitical" civic and commercial leaders, a senate committee killed the Ellsworth bill.[33]

New York's rapid transit commissions pursued ambiguous policies. The board of 1891, even while preparing contracts for a private subway, continued to negotiate with the Manhattan Elevated Company which wanted to extend its old lines as well as build new roads. The rapid transit board created in 1894 disregarded the wishes of the people and sought legislation enabling it to negotiate for a privately owned road. "The proposed underground railway is a work of such peculiar character, and of such exceptional value to the city, that a departure from the settled policy of recent legislation might be justified," the commission rationalized.[34] Even more inexcusable, the board seriously entertained a proposal from the Metropolitan Street Railway Company to build a subway if granted a perpetual franchise. Whatever the rapid transit commission's motives, these activities underscored a lack of ideological commitment to municipal ownership or franchise reform. In this the commissioners reflected the sentiments among an important segment of the utility reform leadership in New York City.[35]

Moreover, few reformers interested in utilities or rapid transit commissioners thought much beyond the construction of a specific project. The rapid transit boards, from those temporary ones created under the act of 1875 to the permanent self-perpetuating one named in the act of 1894, were little more than supervisory construction agencies. Their primary task was the promotion and not the regulation of a transportation system. What restrictions the commissions imposed upon the utility companies were frozen at the time contracts were negotiated; the transit commissions could not exercise continuous and flexible supervision over rates and services. Besides, after the first subway had been built the operating companies were under no obligation "to construct or operate any extensions of the original line or to exchange transfers with the city or any other company operating extensions or connecting lines."[36] By the first decade of the twentieth century, New York City exercised little more control over its transit facilities than it did in 1875 when the first rapid transit act was passed.

Because concerned citizens did not think to involve government on a continuous basis, few considered creating regulatory machinery that would become a part of New York's government. Just as it was assumed reputable men, occasionally coming together to fight against Tammany, could restore clean and honest government, so it was assumed that men of means acting cooperatively could provide for New York's transportation needs. On the day the first train rolled, John Starin of the rapid transit commission gave expression to this faith. "In the midst of . . . rejoicing, let us not forget the men who fought the battle and won the victory," he declared. "Let us remember McDonald the contractor, and Belmont the capitalist, and William Barclay Parsons the engineer, and above all and beyond all let us remember Alexander E. Orr, the president of the Rapid Transit Commission."[37] Few who were rejoicing at the jubilant ceremonies realized the price the city had paid and foresaw that more than seemingly honorable and well-intentioned men would be needed to provide the public adequate transportation services and facilities.

Several decades before the Civil War, gas for light and heat became an important utility in New York City. Early gas franchises contained provisions regulating the breaking up of streets and the quality of gas and construction materials used in the distribution system. In addition, the state legislature in 1859 passed a law creating the office of inspector of gas meters and located it in New York City. Notwithstanding the franchise regulations and lone meter inspector, gas consumers chronically complained about poor service and high rates. What they felt in their purses and saw--rather dimly--in their lamps was substantiated in 1869 by the findings of an investigating committee of the New York State Assembly.[38] With surprising foresight this committee advised, to no avail, that "the entire system of lighting the city should be under the control of practical chemists, whose duty it should be to determine the quality of gas furnished, and, in case the companies do not comply with the law, should have the power to compel them to do so."[39] Like other pioneers, this committee blazed the way for a host of followers. Its success in accumulating accurate financial data helped transform the legislative investigation, often used only as a political hatchet, into an important instrument for reform.

In the mid-eighties when the public again complained about high prices and poor service, and focused this discontent on the newly merged but highly overcapitalized Consolidated Gas Company, a legislative investigation again was conducted. Not surprisingly, a senate committee found in 1885 that the office of gas meter inspector was "useless and afforded little, if any beneficial protection to consumers." The inspector had no power to enforce his findings, and the gas companies could tamper with the meters without removing the inspector's seal.[40] To combat high rates, the legislature upon the committee's recommendation passed a bill limiting the price of gas to $1.25 per thousand cubic feet. Not content with relying on its own regulatory powers, however, and contrary to the wishes of the investigating committee, the legislature also sought to encourage competition by making it easier for gas companies to incorporate.

More significant for later progressive proposals, the committee also recommended and had passed by the legislature a bill to create a board of lighting commissioners to control the gas companies of the city. The commission of three members was granted power to regulate the financial structure and practices of the companies and to force publicity of annual reports showing earnings, disbursements, and amount of gas manufactured and sold. Further, it could supervise the conduct of such companies, investigate consumer complaints, reduce prices if warranted, and regulate the power, purity, and pressure of gas. The commission shared with the legislature the power to allow a company to lay gas mains in the city's streets.[41]

This bill was vetoed by Governor David B. Hill "on the ground both that the powers conferred on the commission were too drastic and that if such a commission were created its jurisdiction should cover the entire state." In his next two annual messages to the legislature, Hill recommended the creation of a state commission "with power under reasonable restrictions to regulate and control the management of all gas companies, throughout the state, such commission to be maintained at the expense of the gas companies." In 1889 and again in 1891, he repeated his recommendation but expanded the commission's jurisdiction to include gas, electric light, telephone, and telegraph companies. Both tries were unsuccessful. Ten years later Governor Benjamin B. Odell recommended that the state's railroad commission regulate gas and electric companies, but no action was taken by the legislature. In 1901 control of gas and electric utility companies by a permanent state commission was still only a suggestion. And it was conceived of primarily as a device to get lower utility rates.[42]

From the mid-nineties to 1905, the battle for cheaper gas and electricity was joined by an assortment of city groups. Businessmen as well as householders desired cheaper utility rates. In fact, business could not function without utility services, and thus many businessmen and commercial organizations were in the front lines battling for additional services and legislative control of utility corporations. In 1897 one businessman asked the Chamber of Commerce to investigate the high rates and, if necessary, recommend restrictive legislation. The burden of high utility costs on New York's poor also placed Tammany Hall, organized labor, and much of the press on the cheaper gas and electricity bandwagon. In keeping with the temper of the nineties, this strange and often incompatible reform army of businessmen, Tammany politicians, muckraking newspaper editors, and labor leaders, isolated a villain in the guise of the gas trust, aroused public opinion, kept the issue on the front pages of the newspapers for a decade, and produced few facts and concrete achievements. The specific, immediate, and politically attractive goal of cheaper rates and better service took precedence over efforts to create an effective method for regulating utility practices.

The main bout centered on changes in gas rates by legislative fiat. A rash of bills was introduced each year after 1897. The most significant and widely supported was that of Democratic Senator Jacob A. Cantor to reduce the price of gas from $1.25 to $1.00 per thousand cubic feet, fix the minimum candle power at twenty-four, and create a permanent municipal gas commission. The $1.00 gas provision caught the popular imagination and, with a mayoral election approaching, Tammany adopted it as a useful political slogan. Also, Tammany had little fear of a local commission. The gas companies recognized the intensity of public feeling by not directly fighting gas reductions. Instead they pushed a compromise bill, written by the Senate Committee on Miscellaneous Corporations, that provided for a gradual reduction of five cents per year until the $1.00 figure was reached. It also required the gas companies to reduce the price of gas supplied to the city from $1.00 to ninety cents. Significantly, the senate committee's bill did not provide for a commission. Other bills providing for municipal ownership of gas utilities were introduced as substitutes for the graduated gas bill but were promptly beaten down by Thomas Platt and the Republicans in the senate. Despite strong opposition from New York City's Democratic senators and Board of Aldermen, the Republicans succeeded in getting the graduated gas bill through the legislature but failed to stop the crusade for cheaper gas.[43]

Over the next several years city legislators, even Republicans, continued to introduce bills for cheaper gas, but the significant debate shifted to the question of the city manufacturing gas and electricity for its own use. Tammany's Croker "is out flat-footed for municipal manufacture and sale of gas," wrote the New York World in 1899. Croker had a bill introduced in the legislature that authorized New York City to build and operate its own gas plant. But with Republican opposition it died in committee. The slippery sachem of Tammany, suggested Pulitzer's paper, introduced the bill to defy the Consolidated Gas Company and embarrass its senate spokesman Boss Platt. No other municipal light plant bill received serious consideration until 1903 when Mayor Seth Low openly challenged the gas monopoly.[44]

New York City's charter provided that public lighting contracts be let after competitive bidding. When Mayor Low took office in 1902, his administration solicited bids for the following year. To the surprise of the gas industry, Low directed Robert Grier Monroe, commissioner of the Department of Water Supply, Gas, and Electricity, to examine and compare the prices submitted. The mayor considered the lighting companies, which occupied public land, as quasi-public corporations not entitled to enormous profits. The city administration also hired an expert to report on the cost of building a municipal electric lighting plant, and Monroe prepared enabling legislation.

No socialistic blueprint, the Low administration's municipal lighting plant bill, introduced in the legislature in March by Senator Nathaniel Elsberg, was limited in application to public lighting needs only. It allowed the Board of Estimate and Apportionment, at its discretion, to authorize the commissioner of water supply to establish a plant to supply electric light to all public streets, parks, buildings, and offices. Naturally, the gas companies fought even this toe in the door of municipal ownership. However, the Citizens' Union, Merchants' Association, Central Federation of Labor, Knights of Labor, and the Brooklyn Young Men's Republican Club sent representatives to Albany to lobby for the bill.[45]

The Merchants' Association took the lead among New York's civic and commercial organizations in rallying strong support for the Elsberg bill. In addition to lobbying, the association published pamphlets attacking the lighting monopoly and its extortionate charges.[46] Before a hearing of the legislature J. Hampden Dougherty explained that the merchants' group was a conservative body that did not "consider the so-called municipalization of public utilities as the panacea for all municipal ills." However, he assured the committee, the merchants abhorred the absence of competition which forced the city to pay a gas price "dictated solely by corporate greed." In the merchants' opinion, the Elsberg bill was a moderate remedy "to rescue the city from the domination of an all-powerful light monopoly which has suppressed competition."[47] From Dougherty's testimony, which was representative of the thinking of a large segment of the business community, it is clear that the businessmen did not consider the city supplying itself with light a step toward socialized utility services. That was simply the means to restore the semblance of competition to the private gas industry, to establish a yardstick against which to measure private rates, and to force down prices. Competition, even with government as the chief competitor, was still thought an effective means to regulate the utilities industry. The bill was killed in the legislature, however. Low lost his bid for re-election to George B. McClellan, and in 1904 John T. Oakly, the new mayor's commissioner of water supply, gas, and electricity, accepted the gas monopoly's offer.

General protest against high gas and electricity rates then focused on the accepted city lighting contracts. In December, Lawrence Veiller wrote Mayor McClellan on behalf of the City Club's board of trustees and asked that the administration investigate Oakly's actions in accepting disadvantageous contracts. In the meantime, the Merchants' Association appointed a committee of five to investigate the lighting trust and prepared a resolution asking the state senate for a legislative inquiry into the contract negotiations. They were supported by the City Club, Citizens' Union, East Side Civic Club, West End Association, the Board of Trade and Transportation, and the Republican county committee.

As the drive for a legislative investigation gained momentum, McClellan's administration renewed efforts to build a municipal lighting plant. After being assured by the corporation counsel that the city had the power under its charter to construct and operate its own lighting plant, the mayor appointed a committee, including Nelsen P. Lewis, the city's chief engineer, and George C. Seavor, an economics professor at Columbia University, to prepare plans, specifications, and an estimate of the cost. To avoid possible legal snags, McClellan had a bill prepared that specifically granted to the city the right to build and operate a lighting plant. In these efforts the mayor received strong support from the Citizens' Union.[48]

In 1905 the legislature was flooded with bills providing for construction of a municipal lighting plant and for reducing gas and electric prices in New York City. Further pressure for utility reform came from the newly organized Municipal Ownership League led by William Randolph Hearst. It was the only organization unequivocally pledged to the public ownership of all utilities. Still, none of the gas bills became law. Instead, the legislature created a joint committee to investigate the entire gas and electric issue in the metropolis. The Merchants' Association, which rightly claimed a large share of the credit for initiating the inquiry, argued that only by exposure of the gas monopoly would private consumers get lower gas prices; this action, they explained, naturally complemented the mayor's efforts to reduce public lighting costs by constructing a municipal light plant.[49] Clearly, the city's merchants had no vision of socialized utility services for private consumers. In fact, Francis B. Thurber and P. S. Tilden of the Board of Trade and Transportation's City Affairs Committee were skeptical even of this limited city venture into the lighting business. "We might better endure the ills we have," they lamented, "than fly to others we know not of!"[50]

For most utility reformers, especially businessmen, who would not follow the more radical Hearst, the alternatives to the prevailing ills were legislative exposes followed by reduced rates or limited city production of gas or electricity, not as a replacement for private enterprise but as a competitive yardstick. In either case the ultimate goal was the same--cheaper gas and electricity prices. Beyond that immediate need, little thought was given to devising a more permanent and equitable method for adjusting interests in the gas and electric business. That awaited a new breed of reformer.

NOTES

1. Delos F. Wilcox, The American City: A Problem in Democracy (New York, 1904), p. 33.

2. William Anderson, The Work of Public Service Commissions with Special Reference to the New York Commissions (Minneapolis, 1913), pp. 6-8, 11; William E. Mosher, "Public Utilities and Their Early Regulation," in History of the State of New York, Vol. VIII: Wealth and Commonwealth, ed. Alexander E. Flick (New York, 1935), pp. 203, 206-211, 233-239; Delos F. Wilcox, Municipal Franchises (2 vols.; Rochester, N.Y., 1910), I, 2-7, 209, 544, 551, 555; Harry James Carman, The Street Surface Railway Franchises of New York City (New York, 1919), pp. 17, 204-220, 235.

3. James B. Walker, Fifty Years of Rapid Transit, 1864-1917 (New York, 1918), pp. 11-60, 126-127; "Annual Report of the Public Service Commission, First District, 1907," New York State Senate Documents, No. 20, IV (131 Sess.), 1908, p. 457 (hereafter cited "PSC Report").

4. Quoted in "PSC Report, 1907," pp. 454-455.

5. Ibid., pp. 455-456; Wilcox, Municipal Franchises, II, 433-438; Chamber of Commerce of the State of New York, Rapid Transit in New York City and in Other Great Cities (New York, 1905), pp. 44-48, 51; Walker, Fifty Years, pp. 59-60, 71, 77, 80-81, 105-106; "Report of the Committee on Claims," New York State Senate Documents, No. 40, V (115 Sess.), 1892, pp. 1-28 (hereafter cited "Report of Claims Committee").

6. "PSC Report, 1907," p. 457; Walker, Fifty Years, pp. 126-127.

7. "PSC Report, 1907," p. 458; Wilcox, Municipal Franchises, II, 439-442; Walker, Fifty Years, p. 107.

8. "PSC Report, 1907," pp. 458-462; Chamber of Com-

merce, Rapid Transit, pp. 51-52; Wilcox, Municipal Franchises, II, 439-445.

9. Allan Nevins, Abram S. Hewitt with Some Account of Peter Cooper (New York, 1935), pp. 497-498, 528-529.

10. Ibid., pp. 498-504; Chamber of Commerce, Rapid Transit, pp. 54-56, 67-68, 80-81; "PSC Report, 1907," pp. 47, 437, 461-463; Walker, Fifty Years, pp. 130-138; Henry C. Wright, "Development of Transit Control in New York City," American Academy of Political and Social Sciences Annals, XXXI (May, 1908), 558-559.

11. Quoted in Chamber of Commerce, Rapid Transit, p. 57; Annual Report of the Corporation of the Chamber of Commerce of the State of New York, 1890-1891, p. 117, 1891-1892, pp. 6, 18-19 (hereafter cited CC Annual Report).

12. New York World, January 4, 5, 6, 7, 9, 13, 18, 28, February 1, 2, 5, 6, 8, 11, 14, March 23, June 20, 1893; The Real Estate Record and Guide, January 7, 1893, pp. 1-6, January 21, 1893, p. 81; Chamber of Commerce, Rapid Transit, pp. 56-60; "PSC Report, 1907," pp. 462-463; Minutes of the New York Board of Trade and Transportation [n.d.], 1893 (hereafter cited BTT Minutes).

13. "PSC Report, 1907," p. 463; Chamber of Commerce, Rapid Transit, pp. 61-65; CC Annual Report, 1893-1894, pp. 89, 92-93, 96.

14. Quoted in Chamber of Commerce, Rapid Transit, p. 66.

15. Ibid., pp. 69-73, 75, 279; CC Annual Report, 1893-1894, pp. xviii-xix, 112-125, 135-137, 1894-1895, pp. 6, 35; Harper's Weekly, XXXVIII (April 28, 1894), 386-387, (May 12, 1894), 435; BTT Minutes, March 7, May 9, 1894; Nevins, Abram Hewitt, pp. 569-572; "PSC Report, 1907," p. 463; Wright, "Development of Transit Control," p. 559.

16. Quoted in "PSC Report, 1907," p. 463; Chamber of Commerce, Rapid Transit, pp. 73-74.

17. "PSC Report, 1907," pp. 50-51, 463.

18. Quoted in Wright, "Development of Transit Control," p. 560.

19. Ibid.; "PSC Report, 1907," pp. 45, 47.

20. Walker, Fifty Years, pp. 141-155; "PSC Report, 1907," pp. 45, 48, 437-439, 463-464; Chamber of Commerce, Rapid Transit, pp. 86-106; BTT Minutes, February 13, 1895; Welling, Diary, May 29, June 3, 5, 1896.

21. Chamber of Commerce, Rapid Transit, pp. 107-109; "PSC Report, 1907," pp. 439, 464; Walker, Fifty Years, p. 155.

22. "PSC Report, 1907," p. 441; Chamber of Commerce, Rapid Transit, pp. 111-120, 153-161; Walker, Fifty Years, pp. 157-161, 163-165, 178-179; CC Annual Report, 1899-1900, p. xiv; Hettrick, "Reminiscences," I, 77-97.

23. New York World, January 4-7, 9, 13, 18, 28, February 1, 2, 5, 6, 8, 11, 14, March 23, June 20, 1893; The Real Estate Record and Guide, January 7, 1893, pp. 1-6, January 21, 1893, p. 81; Chamber of Commerce, Rapid Transit, pp. 56-60; "PSC Report, 1907," pp. 462-463; BTT Minutes [n.d.], 1893, March 8, 22, 1893.

24. Chamber of Commerce, Rapid Transit, pp. 62-65; CC Annual Report, 1893-1894, pp. 92-93, 96.

25. Chamber of Commerce, Rapid Transit, pp. 69-73, 75, 279; CC Annual Report, 1893-1894, pp. xviii-xix, 112-125, 135-137, 1894-1895, pp. 6, 35; Harper's Weekly, XXXVIII (April 28, 1894), 386-387, (May 12, 1894), 435; BTT Minutes, March 7, May 9, 1894; Nevins, Abram Hewitt, pp. 569-572.

26. BTT Minutes, February 13, 1895, March 8, 1899.

27. Annual Report of the Council of the City Club, 1899, pp. 29-30.

28. BTT Minutes, March 8, 1899; New York World, April 20, May 31, 1899; New York Journal, April 26, 1899, Scrapbook No. 17, 1898-1899, The People's Institute Collection, NYPL; Walker, Fifty Years, pp. 156-157; Chamber of Commerce, Rapid Transit, p. 110.

29. Chamber of Commerce, Rapid Transit, pp. 69-73, 75,

279; "PSC Report, 1907," p. 463; Wright, "Development of Transit Control," p. 559.

30. New York World, January 4, 1898.

31. CC Annual Report, 1897-1898, p. 114.

32. Ibid., pp. 112-113; "Ellsworth Rapid Transit Bill" (editorial), Harper's Weekly, XLII (March 26, 1898), 29; ibid. (April 2, 1898), p. 314; New York World, March 19, 21, 1898.

33. Minutes of Special Meeting of the Central City Committee of the Citizens' Union, March 16, 1898, City Committee, Box K-1, CU Coll.; New York World, March 19, 21, 22, 1898.

34. Quoted in Chamber of Commerce, Rapid Transit, pp. 107-109.

35. Ibid., pp. 108-110; "PSC Report, 1907," pp. 440-441; Walker, Fifty Years, pp. 155-157; New York World, January 25, 1899.

36. Milo R. Maltbie, "Elements of a Constructive Franchise Policy," in The Regulation of Municipal Utilities, ed. Clyde L. King (New York, 1914), p. 147.

37. Chamber of Commerce, Rapid Transit, p. 170.

38. Wilcox, Municipal Franchises, I, 544-558; "PSC Report, 1907," p. 451; Anderson, Public Service Commissions, p. 7.

39. Anderson, Public Service Commissions, p. 12.

40. Quoted in Mosher, "Public Utilities," p. 237.

41. Ibid., pp. 237-238; Anderson, Public Service Commissions, pp. 12-13.

42. "PSC Report, 1907," p. 451.

43. CC Annual Report, 1896-1897, p. 95; New York World, January 7, 10, 13, March 12, April 2, 8, 21, 23, 1897; Proceedings of the Board of Aldermen of the City of New York, 1897, Vol. 225, p. 979; New York State Assembly Journal (120 Sess.), 1897, pp. 26, 42, 338, 3438, 3449, 3589-90, 3592, 3652-53, 3656 (hereafter cited Assembly Journal); New York State Senate Journal (120 Sess.), 1897, pp. 22, 173, 641, 679, 680, 1181-84, 2075 (hereafter cited Senate Journal).

44. Senate Journal (121 Sess.), 1898, pp. 32, 465-66, (122 Sess.), 1899, pp. 1282, 1591-93; Assembly Journal (122 Sess.), 1899, pp. 208, 1763, 2611, 3336-37; New York World, February 4, 16, March 1, December 24, 1898, February 1, 2, April 8, 11, 12, 1899, January 4, March 8, 1900, January 10, 1902.

45. "PSC Report, 1907," p. 451; New York World, January 20, March 14, 17, 24, April 3, 1903; New York Times, March 29, 1903, Scrapbook No. 20, 1902-1904, People's Institute Collection; Robert Grier Monroe, "The Gas, Electric Light, Water and Street Railway Services in New York City," Annals, XXVII (1906), 112-114.

46. Merchants' Association, Bulletin, No. 6 (May, 1903), pp. 43-46, "The Lighting Monopoly of New York City and its Exactions" (pamphlet), April 1, 1903, "An Appeal for Justice and Relief" (pamphlet), April 11, 1903, "For an Investigation of the Conditions Surrounding Gas and Electric Lighting in the City of New York," Memorial to the Legislature of the State of New York, January 11, 1905, pp. 1-6, Misc. Scrapbook, Merchants' Association Collection, New York Commerce and Industry Association Library.

47. Quoted in Merchants' Association, Bulletin, No. 6 (May, 1903), pp. 43-45.

48. New York World, April 22, 1903, December 3, 11, 20, 31, 1904, January 10, 17, 26, February 23, 1905; Annual Report of the Council of the City Club, 1905, pp. 50-55; Monroe, "Gas, Electric Light, Water and Street Railway Services," pp. 114-117; BTT Minutes, January 25, 1905; New York Evening Post, December 8, 1904, New York Post, January 14, 1905, New York Evening Mail and Express, December 12, 1904, Scrapbook No. 196, Papers of Charles E. Hughes, Library of Congress.

49. Annual Report of the Council of the City Club, 1905, pp. 55-56; Robert F. Wesser, "Charles Evans Hughes and the Urban Sources of Political Progressivism," New York Historical Society Quarterly, L (October, 1966), 380; New York Post, March 14, 16, 1905, Scrapbook No. 196, Hughes Papers;

"Investigation of the Conditions Surrounding Gas and Electric Lighting," pp. 9-15, Misc. Scrapbook, Merchants' Association Collection; Merchants' Association of New York, Yearbook, 1906, pp. 17-18, Bulletin, No. 7 (April, 1905), p. 2, ibid.

50. BTT Minutes, January 25, 1905.

PART II

THE ARRIVAL OF PROGRESSIVISM

Emerging almost unnoticed amidst the fanfare of patrician politics were new men who sought to redirect reform goals and tactics. Closely identified with the rising professions in such fields as medicine, accounting, law, administration, and social work or with specialized business functions as retailing, banking, and real estate, they consciously sought to extend to the solving of municipal ills their particular expertise. Urging a shift in reform emphasis from men and measures to methods of administration, these younger reformers--progressives--put faith in research and the devising of a technique of reform that could solve New York's immediate as well as future problems. To advance their professional or business interests, gain a hearing from the city government, and assure a continuous influence, progressives formed new organizations or worked through existing ones. In either case the reformers often combined their drive and know-how with the prestige and financial resources of the older patricians, especially businessmen. They discovered even Tammany occasionally in need of experts.

Chapter 5

GOOD GOVERNMENT AND THE BUREAU OF MUNICIPAL RESEARCH

Mayor Low's defeat in 1903, while disappointing to New York's better citizens, actually increased the opportunities for newer and younger men trained in the universities as public administrators, economists, and political scientists. Sharing a faith in scientific method, they enthusiastically tackled the problems of dishonest and inefficient municipal administration with the special tools of their disciplines. Essentially technicians, these progressives gained the confidence of and influenced older patricians, businessmen, and even Tammany officeholders who realized good methods and techniques of governing could serve many ends, could help fulfill a number of visions.

Research-oriented reformers first gained influence under mugwump auspices. In 1901 Frank Tucker, general agent for the AICP and originator of that organization's efficient financial and budgeting system, sent to R. Fulton Cutting an editorial from the New York Daily Tribune which commented upon the AICP's charge that it could manage a public bathhouse more cheaply than the city and implied that under the watchful eye of a private group perhaps other city departments could also be operated more economically. Tucker then formally proposed that the Chamber of Commerce create a bureau or department of civic affairs. Although for a number of years the Chamber as well as other business and civic groups had pleaded for reform in New York City's business methods, nothing came of Tucker's suggestion.[1]

In the meantime, Cutting, who was president of the

AICP, invited William H. Allen, a field worker for the New Jersey Charities Aid Association, to join the staff of the AICP. Allen had done graduate work in economics at the University of Chicago, studied for a while in Germany, and then earned a Ph.D. at the Wharton School of the University of Pennsylvania where he was influenced by Simon Patten, a staunch advocate of efficiency and economy in government. For his thesis Allen did a study of sanitary administration in Pennsylvania.

Soon after joining the AICP as general agent in 1903, Allen began to conduct that relief agency's business according to Frederick Taylor's principles of scientific management.[2] Allen, "a man of extraordinary mental agility, [and] passionate conviction," quickly convinced Cutting that "attacking men as criminals and looking for crime in government--throwing people out of office because of misgovernment--gave little if any protection at all to the public." To have better government, "the public must know how governing was done and must help those in power use better methods."[3] Allen explained to Cutting that a permanent nonpartisan agency staffed with experts--engineers, lawyers, accountants, physicians, and others with special training in the administration of parks, police, fire, health and civic services--was needed. Such a staff would "apply the process of research, and disinterestedness of research, and the techniques of business management, to public affairs and civic problems."[4] Cutting agreed to support such an effort on an experimental basis for one year at $1,000 per month. The Bureau of City Betterment was organized as a branch of the Citizens' Union and Henry Bruere was chosen to serve as director.

Bruere had become interested in social work and community development while studying under Thorstein Veblen, Wesley Mitchell, and Charles Zueblin at the University of Chicago. After graduating he went to Harvard's law school and soon became active in youth work at the College Settlement in Boston. Seeking ideas for the boys' clubs he was organizing, Bruere decided to visit the New York settlements in 1902.

While spending some time with Robert Hunter and his associates at the University Settlement, Bruere was invited by Stanley McCormick, youngest son of the inventor of the reaper, to organize workers' clubs at his father's Chicago plants. The young Bruere accepted and soon became acquainted with Jane Addams, Graham Taylor, Raymond Robins, and others associated with Chicago's settlements. However, Bruere soon returned to New York looking for wider opportunities in social work. After being introduced by Hunter to Allen, Bruere was put on the staff of the AICP. Under Allen's direction Bruere successfully conducted a series of studies, most notably one on public baths, and thus seemed the logical choice to direct the new Bureau of City Betterment.[5]

The bureau exposed corruption and inefficiency in the city's public bath system and in the office of the borough president of Manhattan. Although disliked by most city officials, the bureau was invited by Dr. Hermann Biggs to analyze and suggest improvements in the health department's records and budgeting system. The young researchers also offered to aid Mayor McClellan's Commission on Financial Administration and Accounting, but commission member Frederick Cleveland, a professor of finance at New York University and former student of Simon Patten, had already employed an outside staff to study the city's tangled business records. However, after conferring with Tucker, Cleveland prepared a prospectus for an independent research bureau, and Allen agreed to raise the needed money. Allen and Tucker wrote a brief for an institute for municipal research and conferred with business and professional people, including Richard Watson Gilder, Seth Low, and Carroll D. Wright, a pioneer in labor statistics. It was Gilder, Bruere later remembered, "who suggested the term Municipal Research so as to sharpen the idea that it was dispassionate, scientific, and removed from partisan intentions, and non-political."

Cutting threw his support behind the institute and convinced Andrew Carnegie and John D. Rockefeller to join him in financing the new endeavor. In 1907 the independent Bureau of Municipal Research was incorporated. Among those serving as trustees were Cutting, Tucker, Allen, Cleveland, Wright, Gilder, George McAneny, secretary of New York's Civil Service Reform League and president of the City Club, and Edwin R. A. Seligman, professor at Columbia University. Allen, Bruere, and Cleveland served as directors and supervised the work of the bureau.[6]

The municipal research idea "was that adequate funds should be provided by citizens independent of all party affiliations or governmental alliance, so that a group of staff experts could be employed to find out what the government was doing, how it was organized, what methods were being used, what results were obtained, and to arrive at conclusions as to how the government might be made a more effective instrument for public service."[7] But "instead of getting out on the house-top and shouting abuse at those in office," the bureau was

> first, to submit a statement of facts, asking the officer to have it checked up to determine whether there was any disagreement with respect to conditions; second, to get before the officer an impartial, dispassionate appraisal; third, to show a disposition to be helpful by submitting, upon request, such constructive suggestions as might seem desirable, with supporting reasons for making changes in organization or methods; fourth, to meet with any and all persons in office or out of office, including, when occasion required, the unofficial leader, or "boss," to talk over the pros and cons of the situation; fifth, not to give publicity to the facts and proposals till after the officer himself had been given an opportunity to make his position known; sixth, to keep entirely out of partisan controversy and campaigning, even when attacked.[8]

In this way the average citizen in favor of better government would learn "that there is a technique of intelligence and a technique of efficiency as far beyond the reach of mere goodness as is business efficiency beyond the reach of mere good intentions."[9]

Bureau reformers envisioned a "new city government" fully responsive to community needs and equipped "to further the health, intelligence and economic capacity of its citizens."[10] Ignoring the many political, social, and economic subunits within Greater New York, they defined the entire city as a democratic community and assumed there existed a harmony of interests among citizens and between citizens and elected officials that allowed the promotion of community welfare.

New Yorkers, as all urban citizens, were divided into three distinct governing groups: electors, representatives, and administrators. Electors exercised popular control through ballots, petitions, and charter grants; representative bodies directed the affairs of municipal government through statutes, ordinances, and budgets; and administrators supervised the details of public business through their power to appoint and remove employees and to introduce more efficient methods within the executive departments. Functionally, urban dwellers were categorized as either doers or determiners. Determiners--the body of electors and the representative council--exercised political power; that is, formulated public policy as it emerged from the community will and elected officers and representatives who best articulated that will. Doers--executive officers and their agents--were the administrators who carried out policy and were responsible for services and results. Because bureau men were convinced the primary needs of government were largely business and thus administrative and nonpolitical, they naturally focused attention on doers and their relation to determiners, the relation between administration and politics.[11]

Both politics and administration needed to be brought

under democratic control, the progressives argued. To explain how, bureau reformers compared a city's electors, council, and executive with the stockholders, board of directors, and operating head of a private corporation. As proprietors or stockholders in a public corporation, urban citizens contributed money for municipal services which in the hands of elected officers became trust funds. These officers--trustees--then were obligated by a law of trusteeship to give an accurate accounting of the supplies and services purchased. Money misused resulted in fewer public facilities and an increase in misery and poverty. It followed logically that at all times municipal citizens needed to be informed about the business of their city, to be mindful of "their own business."

To explain why urban citizens failed to manage their own business, bureau leaders turned to history. Convinced that in their day national characteristics had become essentially urban, they nostalgically described small-town, rural Americans as independent and strongly moral. Demanding few services in common and thus little burdened with civic responsibilities or taxes, country dwellers devoted most of their energy to private business and the struggle against nature. Only periodically was it necessary for a community to gather, hear reports from those entrusted with the expenditures of common revenues, and vote for new services and taxes. In a simple, face-to-face manner these rural citizens controlled the business of their governments.[12]

When commerce and industry replaced agriculture as the dominant economic force and expanded occupational alternatives, population increasingly centered around industrial complexes. In these urban centers citizens necessarily were more dependent upon each other and government for services. Yet the pursuit of private gain kept citizen interest in government at a minimum. Public business became more complex as it increased and broadened. Face-to-face observation and a commonly accepted moral code no longer were viable means for intelligent citizen control over government. For want of information, urban dwellers now were "unable to determine at what points waste had occurred or to fix responsibility for loss." Official incompetence and ineffective citizenship were due "to lack of the institutional means" for getting adequate information.[13] When cities were misgoverned, bureau men argued, it was senseless to make public officials "the burnt offering on the altar of public opinion at each election." They should not be held responsible "for the antiquated, ill-adjusted, chaotic, wasteful organization and customary practices of office" they inherited.[14] "In the past," wrote Cleveland, "it has been assumed that the charter powers and 'good' men were the only elements essential to the conduct of the business of a great public corporation. The army of employees, the methods and equipment needed to make this army efficient, have been lost sight of." Urban citizens, then, not only had to retain legal powers of control over elected officials but had to find ways to exercise "intelligent supervision over the administrative acts of . . . [their] representatives and agents."[15]

Therefore, an organized "citizens' agency" was needed to provide citizens the necessary information to make their powers of control effective, explained Cleveland. In fact, he wrote, the single motive of the Bureau of Municipal Research was "to provide an effective agency of citizen inquiry--to make this information obtained concerning the great business in which citizens are interested available for intelligent cooperation." Into the bureau would come data about public business which would then be digested and fed to the public. From the bureau would go suggestions on ways to improve public administration. "It is hoped," Cleveland wrote, "that the intelligence gained through systematic and independent citizen inquiry might be used by the mayor, the comptroller, and heads of departments to correct abuses concerning which a citizen might complain. . . ." The Bureau of Municipal Research, then, both one with and yet apart from "the people," would unite the citizen and his government, provide a means for citizen control over every facet of that government and reconcile expertise with democratic ideology.[16]

The Bureau of Municipal Research could only be as effective as the data it accumulated and New York's archaic government structure allowed. The citizens' agency needed a standard for judging the effectiveness and responsiveness of New York's public officials. What were believed pervasive business principles--economy, efficiency, fidelity--seemed a useful criterion. By fidelity the reformers meant that because a public office was a public trust, elected officials were obligated to account for the properties entrusted to them and for the liabilities they incurred. By efficiency and economy, never clearly distinguished, reformers meant the obligation of officials to render the largest possible service at the least possible cost. For ways to bring municipal government practices into line with these principles the researchers again looked to the private corporation with its centralized authority and systematized bookkeeping.[17]

The staff of the bureau made first-hand observations of methods that seemed successful in the commercial world and appeared adaptable to municipal and state government. Efficiency and management systems employed by private corporations were observed; purchasing methods of several railroads were studied; and the salary schedules in department stores and express, telegraph, and telephone companies were compared with those of public corporations. Bureau men were especially impressed with the increasing importance of staff agencies in private businesses. Unlike a "line agency" which was actually responsible for doing work, a "staff agency" was an independent, critical reviewing body. In private businesses such agencies kept corporation executives and engineers up to date on the latest scientific discoveries and manufacturing processes. Similar organizations in municipal government could help rationalize administration. In fact, bureau men pushed for the employment of more experts of all types in various city departments and bureaus.[18]

After comparing municipal with private administrative

and business practices, the reformers were appalled at the weakness of the city's executive branch. Unlike the president in a private corporation who was solely responsible for the direction of the many employees cooperating for common ends and accountable for the operative results of the business, New York's mayor was not alone responsible for the administration of city affairs. He shared that power with other elected officers, such as borough presidents, the comptroller, and the president of the Board of Aldermen, and with a Board of Estimate and Apportionment that controlled the budget, apportioned taxes, and made contracts with private companies doing work for the city. Executive power had been divided, bureau men explained, to prevent fraud and the concentration of power. The current need was for increased administrative efficiency, and this could be accomplished only through responsible and highly centralized control.[19]

Even more noticeable was the absence in New York's government of adequate accounting methods, a budget system, and other scientific methods for securing and reporting administrative data. Indigenous business firms in rural America had been small enough for the proprietor to keep a sharp eye on his supplies and every business transaction. This close, constant inspection--a system of "personal count"--was of course analogous to the face-to-face control of government by citizens. After the large corporation became the dominant form of business organization a proprietor no longer directly supervised each transaction and the system of personal count was replaced by a system of "record account" that provided accurate data about the business. Although municipal affairs likewise had become big business, no similar accounting systems had been installed thus making it impossible for officials responsible for the affairs of the corporation and public properties to perform their duties intelligently. Furthermore, citizens and taxpayers had no basis for judging the fidelity of officeholders or the efficiency and economy of public administration.

An accounting system in government therefore was a

"first premise to municipal reform." It would include official records and reports and monthly statements of revenues, expenses, assets, and liabilities. Such reports, however, revealed only how honest administrators were and whether they kept within authorized appropriations. To judge economy and efficiency it was necessary to determine the annual cost of administration; public service, like a commercial product, needed to be analyzed on a basis of unit cost. For example, Cleveland explained, it was necessary to devise methods to determine the cost of carrying a passenger or ton of freight by ferry from Manhattan to Staten Island. Such a system of cost analysis needed to be installed in each department of government and summaries deposited in a central administrative office. Published reports based on such information would then enable the administration and public to evaluate how economically and efficiently public business was being conducted.[20]

The most useful report would be an annual budget, an instrument of popular control, planning, and publicity. Although public officials needed legislative approval before raising or granting monies, thus insuring a measure of popular control over the purse, a budget forced public corporate agencies to account for past grants and plans for future revenues and expenditures. In the budget-making process, then, administrators and representatives mutually discussed civic accomplishments, new programs of public service, and methods of raising funds. Here was an excellent instrument for democratizing administration, for allowing strict censorship over government by those who paid the bills.[21] In a positive vein, budget-making allowed the people's chosen representatives "to give direction to the whole corporate policy of a municipality."[22] Such planning, however, could be done only "when accounts and reports reflect the results of experience in terms of cost of work done, with an accurate statement of financial conditions." An accounting and budget procedure, then, allowed "planning the work, planning the expenditures to maintain the work, planning the revenues to meet the expenditures, planning to meet outstanding and maturing credit obligations."[23]

For bureau progressives the tools of administration--budgeting, accounting, statistics, rationalized government structure, and clear lines of responsibility--were means in an urban setting "to make democracy a living, vital thing."[24] Only by efficient municipal methods "could progressive social welfare be achieved," wrote Henry Bruere in his New City Government, "and that, so long as government remained inefficient, volunteer and detached effort to remove social handicaps would continue a hopeless task."[25] Not a "penny-pinching proposition at all," the research movement "showed how people could get done what all the time they had really wanted to get done but didn't know how to do," reminisced Allen.[26] These bureau reformers were convinced that in New York City it was worth the "effort to develop the potential forces of city government for harnessing to the work of social betterment."[27]

NOTES

1.. "A National Program To Improve Methods of Government," Municipal Research, No. 71 (March, 1916), p. 1; Jane S. Dahlberg, The New York Bureau of Municipal Research, Pioneer in Government Administration (New York, 1966), pp. 10-11.

2. William H. Allen, "Reminiscences," CUOHC, I, 12-14, 20-21, 39, 44; Henry Bruere, "Reminiscences," CUOHC, p. 20; Norman Gill, Municipal Research Bureaus (Washington, D.C., 1944), pp. 14, 15, n. 9.

3. Bruere, "Reminiscences," p. 20; Allen, "Reminiscences," I, 47, 54, 66-67, 100.

4. Bruere, "Reminiscences," p. 42.

5. Ibid., pp. 3-18; Gill, Municipal Research Bureaus, p. 14; "National Program," Municipal Research, p. 2.

6. Bruere, "Reminiscences," pp. 23-36; "National Program," Municipal Research, p. 2; Allen, "Reminiscences," I, 92-93; "A Proposed Institute for Municipal Research," December, 1906, Papers of George McAneny, Princeton University; Gill, Municipal Research Bureaus, p. 15, n. 11; Dahlberg, Bureau of Municipal Research, pp. 11-16; "Carroll D. Wright," Dictionary of American Biography, ed. Dumas Malone (New York, 1936), XX, 544-545.

7. "National Program," Municipal Research, p. 1.

8. Ibid., pp. 3-4.

9. "Proposed Institute for Municipal Research," McAneny Papers.

10. Henry Bruere, The New City Government (New York, 1912), p. 8.

11. Frederick A. Cleveland and Arthur E. Buck (eds.), The Budget and Responsible Government (New York, 1920), pp. 6-7, 12-15, 38; Frederick A. Cleveland, Chapters on

Municipal Administration and Accounting (New York, 1909), pp. 65-69, 351-357; Bruere, New City Government, pp. 1-2; "Responsible Government," Municipal Research, No. 69 (January, 1916), pp. 7-8; "Budget Legislation in Two States," ibid., No. 70 (February, 1916), p. v; Dwight Waldo, The Administrative State, A Study of the Political Theory of American Public Administrations (New York, 1948), pp. 42-43.

12. Cleveland, Municipal Administration and Accounting, pp. 1, 10-12, 102-103, 346-349, 356.

13. Ibid., pp. 10, 102-104, 351; Bruere, New City Government, pp. v, 103; Cleveland and Buck, Budget and Responsible Government, p. 103.

14. "National Program," Municipal Research, p. 3.

15. Cleveland, Municipal Administration and Accounting, pp. 102, 357.

16. Ibid., pp. 349, 351; "National Program," Municipal Research, p. 10; "Next Step in the Development of a Budget Procedure for the City of Greater New York," ibid., No. 57 (January, 1915), pp. 1-2; Luther Gulick, The National Institute of Public Administration, A Progress Report (New York, 1928), p. 13.

17. Cleveland, Municipal Administration and Accounting, pp. 70-72, 101-107; Gill, Municipal Research Bureaus, p. 35.

18. Waldo, Administrative State, pp. 5-6; Harold Seidman, Investigating Municipal Administration (New York, 1941), p. 3.

19. Cleveland and Buck, Budget and Responsible Government, pp. 103-105; Cleveland, Municipal Administration and Accounting, pp. 62-65, 118; Waldo, Administrative State, p. 36; "Responsible Government," Municipal Research, pp. 22, 24.

20. Cleveland, Municipal Administration and Accounting, pp. 19-23, 54-55, 70-71, 85-90, 103-105, 115-183, 210.

21. Ibid., pp. 50, 67, 73-80; Cleveland and Buck, Budget and Responsible Government, pp. 39-40; "Budget Systems," Municipal Research, No. 62 (June, 1915), p. i; "Responsible

Government," ibid., pp. 9, 38, 43-53; "Budget Legislation," ibid., p. iii-viii.

22. Cleveland, Municipal Administration and Accounting, p. 74.

23. "Some Results and Limitations of Central Financial Control," Municipal Research, No. 81 (January, 1917), p. 6.

24. Allen, "Reminiscences," II, 159; Cleveland and Buck, Budget and Responsible Government, pp. 6-7.

25. Bruere, New City Government, p. 100.

26. Allen, "Reminiscences," II, 159.

27. Bruere, New City Government, p. 100.

Chapter 6

HEALTH, HOUSING, AND THE TENEMENT HOUSE DEPARTMENT

Hermann M. Biggs was a young man in a hurry. He was born of native American parents in an upstate New York village in 1859. Two decades later Biggs entered Cornell University, did three years work in two, and after deciding that in a medical career he could combine "the two passions of knowing and doing," headed for New York City in the fall of 1882 and entered Bellevue Hospital Medical College. After one semester he returned to Cornell and earned a bachelor's degree. The following year Biggs completed his medical studies at the head of his class and was granted a prestigious internship a Bellevue Hospital. Fortunately for Biggs, New York public health, and slum reform, the young doctor's days of training coincided with a revolution in medical knowledge.[1]

Before the late 1870s, infectious diseases were believed caused by poor sanitary conditions or dirt. The environmental sanitary measures taken in New York City as well as in other American and European cities were based on this understanding. In fact, the organization of New York's health department and the many sanitary rules incorporated in the tenement laws were reflections of this state of medical knowledge. After 1870, however, came one discovery after another by men such as Louis Pasteur, Robert Koch, and Ferdinand Cohn that demonstrated the germ or microbial origin of disease. So rapidly were the causative agents of numerous diseases, including typhoid, tuberculosis, cholera, and diphtheria, identified during the eighties that that decade has been labeled "the golden age of bacteriological discovery." Pasteur and others then successfully experimented with methods of immunization. These advances in microbiology and immunology, pregnant with hope for the masses in urban slums, attracted modern apostles.[2]

Fortunately for New Yorkers, this young science of bacteriology was brought to America or at least advocated by a pioneer group of young doctors--T. Mitchell Prudden, William Henry Welch, Edward C. Janeway, George M. Sternberg, William T. Lusk--many of whom were physicians in New York and on the teaching staff under which Hermann Biggs studied medicine. Almost immediately Biggs himself became a champion of the new science and its application to problems of public health. In fact, his graduating thesis at Cornell after only one semester in medical school was a study of "Sanitary Regulations and the Duty of the State in Regard to Public Hygiene." After pleading for improved sanitary conditions--"pure air and pure water in abundance, sufficiently commodious dwellings and most of all cleanliness"--Biggs called for more study of the causes of disease and hailed the new discoveries of Koch and Pasteur. This thesis, Biggs' biographer notes, was "to be the platform for his life work."[3]

Following a year's study in Germany, Biggs in 1885 was appointed an instructor in the newly built Carnegie Laboratory, the first laboratory building "erected in this country to house one of the medical sciences." Ostensibly under the direction of Drs. Janeway and F. S. Dennis of Bellevue, the new laboratory was actually run by Biggs and became an important center in the application of bacteriology to problems of public health. The young physician soon also served as a pathologist to the City and Bellevue Hospitals, held three teaching positions at the medical school, was a visiting physician for the Workhouse, Almshouse, and Bellevue, engaged in private medical practice, and wrote and delivered several scientific papers.[4]

In 1887 when cholera was suspected on several ships loaded with Italian immigrants in the port of New York, Biggs, Prudden, who worked so closely with Biggs over the next several decades, and a marine hospital surgeon made bacteriological tests on the cultures of the intestinal contents of several deceased immigrants and for the first time in the United States, by biological examination, positively identified the exclusive presence of the cholera spirillum. Here was the beginning of a new technique that heralded the coming of a scientific-sanitary program; and for the next several years Biggs hounded Commissioner of Health Joseph D. Bryant to press for the creation of a division of bacteriology and disinfection within the health department. Not until 1892 did it become a reality. In the interim, Drs. Janeway, Prudden, Biggs, and Horace P. Loomis were appointed consulting pathologists to the Department of Health.

These energetic doctors were not men to wait until consulted. They convinced the department to allow them to formulate a comprehensive report on the cause and contagiousness of tuberculosis and to recommend to the city specific measures that would protect New Yorkers from the dread disease's influences. Because many of the city's doctors were skeptical of the judgments of the report, the health board would only support an educational program informing Gothamites of practical sanitary measures to prevent consumption. Still, this action in 1889 probably was "the first movement officially taken by any municipal sanitary board toward the formal sanitary supervision of tuberculosis.[5]

By this time Biggs and Prudden were dreaming of the wider "application of the new science of bacteriology to the practical problem of the public health."[6] In 1892 their chance came when a cholera epidemic hit Hamburg, Germany, the departure point for many immigrants bound for New York. The Chamber of Commerce appointed an advisory committee on cholera of which Biggs was a member and the Board of Health swung into action to clean up the city. The water supply was inspected and the streets thoroughly cleaned; tenement inspection was increased so as "to strike at the breeding places of the allies of cholera, as well as of cholera itself," and circulars of instruction on the prevention and treatment of cholera were issued.[7]

When the first infected ships began arriving, Biggs and Commissioner Bryant exploited the crisis and convinced the Board of Health to create a new division of pathology, bacteriology and disinfection. Immediately a system of disinfection was organized, a diagnostic laboratory set up, and with the help of Dr. Edward Dunham, who had recently returned from the Hygienic Institute in Germany, and several chemists and technical assistants, the diagnosis of the cholera cases began and the disease successfully contained. "Thanks to Dr. Biggs," writes his biographer, "New York City thus has the credit of establishing the first health department laboratory in the world for the application of this new science to the routine diagnosis of disease--a step which laid the firm basis of fundamental science upon which American public health procedure was thereafter to be built."[8] As public health historian George Rosen has written, "Furthermore, the laboratory represented the practical outcome of the microbiological period, just as the organization of the health department had been a product of the earlier sanitary reform movement. Just as the health department provided an appropriate administrative mechanism for dealing with community health problems, so the public health laboratory provided a suitable scientific tool for the implementation of the public health program."[9]

In this new permanent laboratory over the next several years "the discoveries of Pasteur, Koch, and others were systematically applied to the protection and improvement of the community's health."[10] Diphtheria and then tuberculosis, two dread diseases, were immediately attacked through bacteriological methods. In the war against diphtheria, Biggs, always ready to employ "other men with special knowledge," brought into the laboratory Dr. William H. Park, a young physician who previously had been engaged in similar research. Out of the laboratory's studies came methods to scientifically differentiate between diphtheria and other allied conditions and proof that the organism that caused diphtheria, and by implication the germs that caused other community diseases, could be carried by healthy persons.

When in 1894 diphtheria antitoxin was successfully prepared in Europe, Biggs, in Germany at the time, cabled home the news and ordered Park to begin the process to prepare the serum. He secured the necessary funds first from a subscription drive sponsored by the New York Herald and then from the Board of Estimate. In spite of strong opposition from a nonbelieving segment of the medical profession, Biggs, emerging as a public health administrator and statesman, prepared and had passed by the state legislature a bill authorizing the Department of Health to prepare and sell diphtheria and other antitoxins. By 1896 even most of the opposing doctors were acquiescing.[11]

Soon after the fight against diphtheria had been launched, Biggs turned his attention again to tuberculosis. In a second report on the subject to the Board of Health, Biggs emphasized that the disease could only be acquired by receiving into the body the living germ, but that it was communicable and preventable. Never content just to describe a bad situation, Biggs also outlined a bold program for the restriction and prevention of consumption. As summarized by his biographer, Biggs recommended "a systematic campaign of popular education, the reporting of cases of tuberculosis by public institutions, the assignment of special inspectors to investigate the disease, to educate affected families in its control and management, and to take special measures for disinfection, the inception of routine bacteriological examinations for diagnosis of the disease, and the provision of proper hospital facilities for its care."[12] Biggs also recommended that private physicians be asked voluntarily to notify the Board of Health of all tuberculosis cases under their care. When the board in December, 1893, adopted only part of his program, Biggs the next month sent another plea to the health administrators. In February his entire program "establishing a complete administrative policy" was adopted.[13]

But a major clash with more conservative members of

the medical profession erupted in 1897 when Biggs and Prudden finally got from the Board of Health an ordinance declaring tuberculosis an infectious and communicable disease dangerous to public health, thus requiring by law its reporting by all physicians. It was of a "dictatorial and defiantly compulsory character"[14] and "a mistaken, untimely, irrational and unwise" move, wrote the Medical Record. The Medical Society of the County of New York unanimously declared the compulsory reporting edict "unnecessary, inexpedient and unwise." More threateningly, the county medical societies tried to convince the state legislature to strike those parts of New York City's charter giving the health department power to deal with communicable disease.[15]

Convinced that the power of government and law should be harnessed behind a program to extend scientific knowledge to community health problems, Biggs, Prudden, and their young associates had no aversion to political engagement. Biggs spent the winter of 1898 and 1899 in Albany fighting passage of such bills, and with the Board of Health supporting him, won the skirmish. By 1900 he was convinced that most active opposition had ceased and the city's doctors were cooperating in the battle against tuberculosis, especially among the city's tenement house population.[16]

Firmly embedded within New York City's health department was a permanent agency staffed with young, energetic medical scientists and equipped to tackle in a systematic fashion major urban disease problems. "Everything which is detrimental to health or dangerous to life," declared the overly optimistic Biggs in 1897, "is regarded as coming within the province of the Health Department."[17] Primarily at his initiative this was to approach fulfillment after 1902 when Biggs served as general medical officer for New York City. In the meantime, the health department extended its organization to the newly created Greater New York, began a program to examine sick school children, and engaged in a good deal of educational work, especially in the tenement districts. Biggs took part in efforts to reorganize medical education in New York City and to encourage the establishment of public health as a distinct career in medicine. This young doctor, representative of a new brand of reformer--a specialist, confident his professional skills could be utilized to solve community problems--emerged "as the sanitary statesman."[18] In 1900 he and fellow workers cooperated with Lawrence Veiller and the tenement house commission in exposing the connection between living and working conditions and poor health.

Lawrence Veiller was another young man in a hurry. Though born in New Jersey , he attended City College in New York and after graduating in 1890 plunged into social work on Manhattan's lower East Side. Observing tenement life at its worst, especially during the depression that began in 1893, the precocious Veiller at twenty concluded that improved housing for the poor "was the starting point of everything."[19] Veiller's origin as a housing expert began when he went to work as a construction plan examiner in the Department of Buildings during Mayor Strong's administration.[20]

Not impressed with the limited tenure and accomplishments of New York's several tenement house committees, Veiller in 1898 "conceived the idea of starting a movement which would be permanent and which would continually seek to improve the conditions of the tenement houses." His scheme for the creation of some type of permanent tenement house commission or society was turned down by the AICP and the State Charities Aid Association, but the young reformer gained a hearing from Robert W. deForest, president of the Charity Organization Society. Veiller explained that tenement conditions were deteriorating, existing laws were neither enforced nor adequate, and well-organized builders constantly lobbied to weaken the laws even further. Under authority granted by the new Greater New York Charter, the municipal assembly was shortly to appoint a commission to revise the city's building code.[21] The impatient Veiller pleaded that a new, permanent society could "secure the enforcement of the existing laws, . . . present united opposition to bad legislation, . . . obtain such new or remedial legislation . . . [and make a] general study of the tenement-house question."[22]

DeForest--"his mind always open to new things"--was convinced, and toward the end of the year a tenement house committee was organized within the Charity Organization Society.[23] Included on the committee were the long-active Richard Watson Gilder, Felix Adler, and Jacob A. Riis; deForest and Edward T. Devine, general secretary of the Charity Organization Society; Frederick W. Holls (chairman), George B. Post, Ernest Flagg, and I. N. Phelps Stokes, all architects; E. R. L. Gould of model tenement fame; Constant A. Andrews, active in tenement work with the AICP; and John Vinton Dahlgren, former attorney for the buildings department in New York. Veiller served as secretary and executive officer of the committee and was the guiding spirit behind its activities.[24]

When the Board of Aldermen appointed a commission in January, 1899, to revise New York's building code, Veiller, working through the new tenement committee, proposed fifteen ordinances which would improve fire, ventilation, and sanitary facilities, rallied reformers behind them, and publicized their importance. Although these recommendations were supported by the New York Chapter of the American Institute of Architects, the Architectural League, the AICP, the Children's Aid Society, various settlement houses, the City Club, the Social Reform Club, the Citizens' Union, and even several state and city officials, none was adopted by the Tammany administration. In fact, the new code was a definite step back "toward the slough out of which it was pulled with such toilsome effort." Now tenements could be built higher and could contain inflammable wooden dumbwaiter shafts, and wooden houses could be occupied by six rather than only three families.[25]

The tenement reform cause did not collapse with this defeat. A permanent committee, staffed with housing authorities and led by a specialist whose full-time job was housing reform, now existed. Here were no "queen hornets" or "mornin' glories." To regain the initiative, Veiller, as he recalled many years later, "conceived the happy idea of bringing the so-called slums to the people uptown, instead of trying to take the uptown people down to the slums."[26] Working with the Charity Organization Society's housing committee, he organized in February, 1900, a tenement house exhibit that allowed New Yorkers to view for two weeks cardboard models, photographs, maps, charts, and diagrams of existing blocks of slum houses and model tenements, lodging houses, and hotels. Statistical tables and maps revealed the density of population, death rates, ethnic composition, and amount of poverty and disease in the tenement districts. Such detail, Veiller suggested in his history of the event, "might well earn for New York City the title of the City of Living Death." In conjunction with the exhibit, conferences were held where leading specialists discussed various phases of tenement problems. "The Tenements and Tuberculosis," "The Tenements and Poverty," and "Improving Tenements by Personal Influence" are suggestive of the range of talks. Also, over 170 architects competed in a model tenement plan contest.[27]

Governor Theodore Roosevelt was one of the ten thousand or so that viewed the exhibit, and he was especially impressed with the poverty and disease maps. Only a few years earlier as a young assemblyman, he had toured the tenement districts with Samuel Gompers and supported labor's drive to outlaw the manufacture of cigars in tenements. As New York's controversial and peripatetic police commissioner under Mayor Strong, and as a close friend of the poor-man's advocate Jacob Riis, Roosevelt had gained further insight into slum life. To the delight of the housing reformers, Roosevelt promised deForest help in securing remedial legislation. "We must have intelligent legislation to penalize the doings of those who cannot be reached by the motives of humanity," he told an audience at the exhibit.[28] As governor, Roosevelt often sought the aid of experts in fashioning a legislative program. This opened the door to achievement for housing progressives, but more significantly, it presaged a functional relation between political decision-makers and progressives.

After consulting with deForest, Veiller prepared a bill authorizing the governor to appoint a state tenement commission and journeyed to Albany to lobby for its passage. Opposed by Tammany and the building "ring" in New York City, the bill was not reported by an assembly committee until Roosevelt, pressured by Veiller, pleaded with Jotham P. Alds, the committee's chairman. "Oh Lord," the governor wrote, "I wish it could pass!" The bill finally got through the assembly but was stalled in the senate. To avoid a possible veto by Tammany Mayor Van Wyck, who under the constitution had to approve this special legislation for New York City, Veiller prepared and had his friends in the Charity Organization Society of Buffalo introduce an amendment to his bill widening its application from New York City alone to all cities of the first class, thus making it a general bill. In this form and with Roosevelt coaxing the Republican leadership, the bill creating New York's tenement house commission of 1900 was passed and promptly signed by the governor.

Given a free hand by Thomas C. Platt and the Republican organization, Roosevelt, in consultation with deForest and Veiller, chose the fifteen members of the new commission. DeForest was elected chairman and Veiller appointed secretary.[29] What had been a permanently organized voluntary cooperative effort for housing reform had now harnessed the power of government to its efforts.

More than a secretary, Veiller stamped his image on the entire proceeding. A comparison of the commission's final reports, of which Veiller had a large share in writing, and a proposed plan of work prepared by him in June, 1900, before the commission began its inquiry bear remarkable similarities. In fact, this professional housing reformer brilliantly and systematically directed the commission's work, prepared many of the special reports, secured the aid of extraordinarily talented men, wrote most of the commission's final reports and the tenement house law of 1901, and planned the permanent state tenement house department.[30]

Like the small band of doctor-reformers shaping New York's public health program, the tenement house commission relied on "expert information and expert opinion." Many specialists, including Drs. Biggs and Arthur R. Guerard, conducted for the commission selective investigations and prepared written reports. Information and advice were secured from the heads of the fire, buildings, and health departments, architects, engineers, builders, and even tenement dwellers. Veiller, though not one to minimize his own achievements, recognized his debt to earlier tenement house investigations and laws and in a magnificent piece of research codified all existing laws relative to tenements and wrote a history of previous reform efforts. Studies also were made of housing conditions and tenement laws in other American and European cities.[31]

Thus armed with a wealth of data and historical perspective, Veiller and the commissioners were painfully aware that in spite of a half-century of reform effort conditions had worsened. By 1900 over two-thirds of Gotham's citizens lived in legally defined tenement houses. "With all the remedial legislation and regulation which has been put into operation since the enactment of the first tenement house law in 1867," reported the commission, "the present type of tenement house--the six-story double-decker--occupying 75 per cent of a 25-foot lot, with four families on a floor, gives to its occupants less light and less ventilation, less fire protection and less comfortable surroundings, than the average tenement of fifty years ago, which was lower in height, occupied less lot space, and sheltered fewer people."[32] Slum houses were "centers of disease, poverty, vice, and crime" and thus dangerous to the welfare of the community, implored the commissioners in a tone not unlike that of their predecessors.[33]

The previously established rules of the tenement game

set some outer bounds of decency but still were rigged in favor of the builder, speculator, and owner. The laws were not changed frequently enough to keep "pace with the conditions sought to be remedied," and rarely was there an umpire around to enforce them.[34] To prevent bad housing in the future and correct tenement evils of the past, Veiller and his co-workers realized a new and better rule book was needed. Thus, like earlier commissions, they recommended a tightening of the code regulating the fireproofing, sanitation, and ventilation of tenements. This time the reformers buttressed their arguments with the reports and opinions of New York's leading physicians armed with the germ theory of disease and special studies made by civil and sanitary engineers and other experts.

The relation between poor housing and disease, previously grounded in the filth theory of disease and responsible early in the nineteenth century for the joining of the health and housing reform campaigns, was not scientifically established. "Experimental investigations have shown clearly enough that the tubercle bacillus--the only necessary factor in the production of tuberculosis--is readily destroyed by sunlight . . . and in this, as in all other communicable diseases, the danger of infection is largely diminished by thorough ventilation," wrote Dr. Biggs in a report for the tenement house commission. "But in too many of the houses in New York City there is never sufficiently thorough cleaning to remove the infection, and sunlight, daylight, and fresh air sufficient to destroy the tubercle bacilli cannot enter."[35] Both Drs. Biggs and Guerard, in separate studies, called for sanitary and ventilation improvements that found substance in the commission's major report and revised code.[36]

New tenements were to have large courts proportionate in size to the height of the building, thus eliminating the narrow dumbbell air shaft--"a stagnant well of foul air"; bathrooms were to be installed in every apartment; public hallways on each floor were to have windows opening to the outer air; nonfireproof buildings less than forty feet wide could not be built beyond five stories, and a fireproof floor had to separate the cellar from the rest of the building; public halls and stairs, as under previous laws, had to be completely fireproofed. Existing houses were to be remodeled so as to provide in dark, interior rooms large windows opening into outer rooms with fresh air and light; wooden panels on doorways in public hallways were to be replaced with glass; sewer-connected privy or school sinks in the yards were to be replaced by modern water closets on each floor of the house, and better fire escapes were to be intalled.[37] These changes, especially in regard to new tenements, were a significant advance in restrictive housing legislation. That subsequent tenements were referred to as "new law" in contrast to "old law" tenements was indicative of the "virtual revolution in New York's housing conditions" that the Tenement House Law of 1901 began.[38]

The fatal flaw in a half-century's efforts at reform, however, was a "lack of continuous effort." As complicated a social problem as slum housing required "constant and continuous attention."[39] Continuity and regularity--the heart of progressive concern--in the city's management of pressing urban problems demanded more than a better rule book. "With an efficient system of administration and enforcement of the tenement house laws much could be done to remedy many of the evils that exist," suggested Veiller.[40] Increasing the number of inspectors in the health department, as previous reports had urged, was insufficient because the department had too many other responsibilities and had no bureau concerned only with tenements. Besides, the health department shared responsibility for tenement supervision with the buildings, fire, and police departments, each of which had other major concerns. With responsibility so divided, enforcement of the tenement code was virtually nonexistent, and, moreover, no one person or department could be held accountable.[41]

Only "the continuous attention, year in and year out, of some one body directly charged with the enforcement of tenement house laws" would assure their proper administration.

The best way to achieve this and solve "the present and future tenement house problem," concluded Veiller in the commission's report, was "through the creation of a separate department charged with no duty except the supervision of the tenement houses--a department of such size as to adequately fulfill such functions."[42] So Veiller prepared a bill creating a tenement house department which the state legislature accepted, along with the new tenement code. The new department was incorporated into New York City's charter. The Tenement House Department provided "a wonderful opportunity for those of us who had counted on the enforcement of the new law and the new piece of municipal machinery to bring about great reforms," reminisced the brilliant Veiller some years later.[43]

NOTES

1. C.-E. A. Winslow, The Life of Hermann M. Biggs, Physician and Statesman of the Public Health (Philadelphia, 1929), pp. 1-48, 59.

2. Rosen, History of Public Health, pp. 294-332; C.-E. A. Winslow, The Contribution of Hermann Biggs to Public Health, reprinted by the New York Tuberculosis and Health Association from The American Review of Tuberculosis, XX, NO. 1 (July, 1929), 2-5.

3. Winslow, Life of Biggs, pp. 40-60; Rosen, History of Public Health, p. 332; Donald Fleming, William H. Welch and the Rise of Modern Medicine (Boston, 1954), pp. 57-65.

4. Winslow, Life of Biggs, pp. 60-66, 75.

5. Ibid., pp. 79-87.

6. Ibid., p. 90.

7. Ibid., pp. 93-95.

8. Ibid., pp. 95-97.

9. Rosen, History of Public Health, p. 335.

10. Ibid., p. 333.

11. Winslow, Life of Biggs, pp. 101-130.

12. Ibid., pp. 132-133; Winslow, Contribution of Biggs, p. 13.

13. Winslow, Contribution of Biggs, p. 13; Winslow, Life of Biggs, pp. 133-134.

14. Quoted in Winslow, Life of Biggs, p. 144.

15. Winslow, Contribution of Biggs, p. 14.

16. Ibid., pp. 14-16.

17. Quoted in ibid., p. 16.

18. Winslow, Life of Biggs, pp. 122, 151, 153, 161, 166.

19. Lawrence Veiller, "Reminiscences," CUOHC, I, 3.

20. Lawson Purdy, "Reminiscences," CUOHC, p. 35.

21. Veiller, "Reminiscences," I, 5-9.

22. Draft of a proposal "to establish a permanent Tenement House Commission, or Society, as a new philanthropic body," April 26, 1898, pp. 3-4, Papers of Lawrence Veiller, Columbia University.

23. Veiller, "Reminiscences," I, 9.

24. Veiller, "Tenement House Reform," p. 110; Lubove, Progressives and Slums, p. 119.

25. "Tentative Suggestions for Ordinances Supplemental to . . . the Present Tenement House Laws as Embodied in . . . the Charter," proposals enclosed in a brochure dated May 8, 1899, inviting recipient to attend conference sponsored by Charity Organization Society's Tenement House Committee, Veiller Papers; Veiller, "Tenement House Reform," pp. 110-111; Eighteenth Annual Report of the Charity Organization Society of New York City, July, 1899-June, 1900, pp. 13-14; Jacob A. Riis, "The Tenement House Exhibition," Harper's Weekly, XLIV (February 3, 1900), 104.

26. Veiller, "Reminiscences," I, 11.

27. Ibid., pp. 11-15; "Prospectus of Tenement House Exhibition," July 19, 1899, Veiller Papers. This prospectus was sent with a questionnaire on housing to individuals in various cities in the country. Veiller, "Tenement House Reform," pp. 111-116.

28. Quoted in G. Wallace Chessman, Governor Theodore Roosevelt: The Albany Apprenticeship, 1898-1900 (Cambridge, 1965), p. 232; Veiller, "Reminiscences," I, 19-20.

29. Theodore Roosevelt to Jotham P. Alds, undated, Veiller Papers; Chessman, Theodore Roosevelt, pp. 232-233; Veiller, "Reminiscences," I, 20-26; William H. Harbaugh, Power and Responsibility: The Life and Times of Theodore Roosevelt (New York, 1961), pp. 123-124. The commission, in addition to deForest and Veiller, included Otto M. Eidlitz, Myles Tierney, and Alfred T. White, all builders; I. N. Phelps Stokes, Raymond T. Almirall, and William Lansing, architects; lawyers William A. Douglas and Paul D. Cravath; Charles S. Brown, real estate agent and broker; James B. Reynolds, headworker of the

University Settlement; George B. Fowler, former health commissioner; Hugh J. Bonner, ex-chief of the Fire Department; William J. O'Brien, a union official in the building trades; and F. Norton Goddard, a merchant. Jacob A. Riis, "The New York Tenement-House Commission," American Monthly Review of Reviews, XXI (June, 1900), 689-692.

30. These reports constitute deForest and Veiller (eds.), The Tenement House Problem; see draft of Veiller's report to the Executive Committee of the Tenement House Commission, June, 1900, Veiller Papers, for his personal views of the work to be done by the commission. Veiller, "Reminiscences," I, 27-28.

31. Robert W. deForest and Lawrence Veiller, "The Tenement House Problem (Being the general report of the Commission)," in The Tenement House Problem, I, 11, 67.

32. Ibid., pp. 3, 5.

33. Ibid., p. 10.

34. Ibid., p. 5.

35. Biggs, "Tuberculosis and the Tenement House Problem," ibid., I, 447.

36. Ibid., pp. 447-458; Arthur R. Guerard, "The Relation of Tuberculosis to the Tenement House Problem," ibid., I, 361-470.

37. DeForest and Veiller, "Tenement House Problem," pp. 10-24; Robert W. deForest, "Introduction: Tenement Reform in New York Since 1901," in The Tenement House Problem, I, xiv-xix.

38. DeForest, "Introduction," p. xix.

39. DeForest and Veiller, "Tenement House Problem," p. 11.

40. Lawrence Veiller, "The Speculative Building of Tenement Houses," in The Tenement House Problem, I, 376.

41. DeForest and Veiller, "Tenement House Problem," pp. 25-29.

42. Ibid., pp. 29-30.

43. "Skeleton Outline of a Plan for a New Tenement House Department" enclosed with letter from Veiller to

deForest, October 17, 1900, Veiller Papers; Veiller, "Reminiscences," I, 31-36.

Chapter 7

UTILITIES AND THE PUBLIC SERVICE COMMISSION

Progressives concerned with utility reform relied on functional, contractual, and administrative arguments in their fight to use the power of government to force utility corporations to exercise economic and social responsibility. In several influential volumes on municipal government and franchise policies, Delos F. Wilcox cogently argued the public nature of utility services. "The operation of public utilities is public business, and . . . its controlling motive must be service rather than the exploitation of urban necessities."[1] In 1907 Governor Charles E. Hughes reminded a business audience that the privileges utility corporations exercised had "been created by the public for public benefit, and not primarily for private advantage."[2] Appalled at corporate abuses, other progressives insisted that public regulation of utilities "must seek not only adequate service at reasonable rates, but efficient and economical management of corporate affairs as well."[3] For too long utility services, rates, and business practices "were because they were. Like Topsy they had just grown."[4] Progressives hoped to replace chaos with order.

By 1907 New York still possessed only one subway line. "The conditions under which people are transported in New York city are indecent, and the present facilities are wholly inadequate," investigators reported. Although the rapid transit board had outlined nineteen new subway routes, no construction contracts had been awarded.[5] Critics of the commission--Ray Stannard Baker labeled it "a board of fine old gentlemen"--and advocates of better transit service often ignored the commission's lack of regulatory power over all modes of city transportation.[6] The board's supervisory power over the existing subway was negligible because its influence virtually stopped at the print on the contracts. Its power to grant new franchises to existing elevated lines was contingent upon the private company's desires, and the board could not supervise existing services on the elevated and street surface railways, no matter how overcrowded, poor, and inadequate.[7] By law the state board of railroad commissioners, established in 1882 "to be an investigating, reporting, and regulating body," had jurisdiction over the surface and elevated roads in New York City as well as over the state railroads.[8]

Although on paper this permanent state commission had extensive power to regulate services and facilities, it could not effectively enforce its recommendations. That was the responsibility of the state's attorney general, who in no way was obligated to act on the commission's orders. In practice the state board did little more than publish financial reports, but often even these were filled with erroneous information. "The board operated as a protection to the railroads rather than as a corrective body," reported Henry C. Wright in 1908.[9]

The Merchants' Association, backed by several groups of angry women indignant over the abuses women experienced on the crowded transit facilities, recognized the weaknesses of the state railroad commission and the need for more effective public regulation. The merchants organized mass rallies, published pamphlets, conducted investigations, and appealed to the railway companies. "Which Shall Prevail, The Greed of Monopolies or the Rights of the People?" and "An Appeal for Justice and Relief" were the crowd-pleasing titles of two of their pamphlets. In 1903 the association's legal committee drafted and had introduced in the legislature a bill to enlarge the membership of the state board and assure that two members be engineers and local residents of New York City. The orders of the commission were to be honored unless appealed in the courts. The bill failed four consecutive years. However, this agitation did secure a local office of the state railroad commission.[10]

Few reform organizations took part in the Merchants' Association's campaign because its bill violated notions of home rule. More popular was a Board of Trade and Transportation bill to create a local board of railroad commissioners with jurisdiction over all railroads and streetcars in the city and the power to deal with issues beyond the responsibility of the rapid transit commission and too local to attract sufficient attention from the state commission. In 1904 over twenty-one civic and commercial organizations rallied behind this scheme. These efforts met the same fate as those of the Merchants' Association. Yet the Board of Trade and Transportation vowed to continue to fight for a city railroad commission.[11] At a meeting of the board in 1905, Calvin Tomkins declared "the interests of the transportation corporation should be made subservient to the general public interest instead of allowed to dominate it, as has been the case heretofore."[12] The businessmen at least were groping for a more effective means for the city and state governments to promote and regulate much-needed transportation services. The solution was to come from another direction.

In response to the Merchants' Association's fight for cheaper gas and electric rates, the state legislature in 1905 appointed a joint committee headed by Senator Frederick C. Stevens. Charles Evans Hughes, a young, brilliantly successful New York lawyer, was chosen chief counsel. Hughes was relatively unknown outside the legal profession and had no experience in public service. Possessed with superb intellectual abilities, a sense of moral well-being, and an exceptional capacity for work, this self-confident son of a Baptist minister achieved full partnership in a respected law firm by his twenty-fifth birthday in 1887. In the handling of minor legal services for several electric light and wire companies, Hughes became acquainted with James C. Carter, James H. Choate, Elihu Root, Wheeler H. Peckham, and other important members of the

New York bar and leaders in the fusion movements of the nineties. Unlike these older contemporaries, however, Hughes' only connection with the reform politics of that era was his vote for fusion candidates and his frequent appearances at the Social Reform Club. Primarily concerned with his profession, this precocious young lawyer had little time for politics.

In legal style as well as in political temper, Hughes was different than his older and talented colleagues at the bar. Less concerned than they with forensic skills and courtroom dramatics, he put full confidence in research and facts. Blessed with an almost photographic memory for details, Hughes carefully organized his work so that in court he could bury opponents in an avalanche of facts. Early in his career Hughes represented the owners of a Michigan beet-sugar factory in their suit against a German firm which they claimed had built and installed faulty machinery. He visited the plant, made a thorough study of beet-sugar manufacturing and "came back with enough technical details and expert testimony to overwhelm the opposition." Efforts such as this earned Hughes the reputation as a lawyer's lawyer and revealed his kinship with young progressives in New York City. Significantly, at approximately the same time that Henry Bruere, Frederick Cleveland, Lawrence Veiller, John Purroy Mitchel, and other bright young college graduates were seeking through the creation of private research organizations and the utilization of government investigatory powers to extend their skills and concepts of research-oriented reform to municipal problems, New York's state government sought Hughes' legal expertise for a pressing public issue.

Unknown to Hughes, Governor Francis W. Higgins, at the suggestion of former Judge William N. Cohen, recommended him as counsel to the Stevens committee. Stevens was unenthusiastic at first because he hoped to secure a lawyer well known to the public. But after several New York lawyers, including Henry Taft who had opposed Hughes in the beet-sugar case, and newspapermen associated with the New York World spoke highly of Hughes, Stevens offered him the job.

Taft had explained to Stevens that Hughes was one of a few who could cut through the maze of technical testimony that was sure to fill the hearings.[13]

Skeptical of the honesty of legislative inquiries and without political ambitions, Hughes at first declined the offer, claiming he knew nothing about the gas utility business. Only after Stevens assured Hughes that politics would play no part and that he would have a free hand in the investigations did Hughes reconsider and accept the job as counsel. "There is a public service of importance to be rendered," he convinced himself, "and if I go at it with the single purpose of doing a thorough professional job, that attitude will be recognized and in some way I shall get through."[14] Hughes did a professional job. "With the suddenness of an eruption," writes his biographer, "a striking new figure burst in upon the public consciousness." For nine successive weeks Hughes blended brilliant insights, probing questions, and a polite manner to pry loose information from the chief executives of all the gas and electric companies operating in New York City, various city officials, and even Charles F. Murphy, the tight-lipped Tammany boss.[15] "Mr. Hughes has shown an ability for this kind of work such as very few men possess," commented the New York Times as the hearings came to a close.[16]

The facts about the gas and electric business added up to a sordid tale of corporate greed and profiteering, overcapitalization of assets, poor service, and other charter and franchise violations that explained New York's exceptionally high utility rates and poor service. For a week after the end of the hearings, Hughes shaped the data into a report with recommendations for corrective legislation. In school-like fashion he explained the report's contents to the state legislators and helped draft bills to convert its recommendations into law. Of the eight bills introduced by the Stevens committee, only one failed to pass. This would have reduced the price of gas from $1.00 to seventy-five cents per thousand cubic feet. The next year, however, an eighty-cent gas bill passed the legislature.

Bills were passed that regulated gas pressure, purity, illuminating power, and measurement. Electricity was reduced from fifteen to ten cents per kilowatt hour and electric meters were to be periodically inspected. Street arc lamp prices were greatly reduced, and New York was given authority to utilize water power to generate electric current for public purposes.[17]

In his prescient analysis, Hughes rejected both competition and irregular legislation. "The gross abuse of legal privilege in over-capitalization and in the manipulation of securities for the purpose of unifying control and eliminating all possible competition," he wrote in his report, "shows clearly that there can be no effective remedy by general legislation; . . . and that for the protection of the public there should be created a commission with inquisitorial authority, competent to make summary investigation of complaints, to supervise issues of securities and investment in the stocks or bonds of other companies, to regulate rates, and to secure adequate inspection; and otherwise, to enforce the provisions of the law."[18] Hughes helped translate this sentiment into a bill that provided for a state gas and electricity commission, consisting of three members appointed by the governor with the consent of the senate.

The commission was granted power to investigate upon complaint or on its own initiative the quality of service and the methods employed in the manufacture or distribution of gas and electricity. It could also supervise the issuance of stocks and bonds, the transfers of franchises, and the purchases by corporations of stocks or bonds issued by competitors. However, the commission's rulings were not self-enforcing. Court orders were necessary to compel obedience to the commission's decisions. Even more damaging to its effectiveness was that upon appeal to the appellate division of the state supreme court the commission's orders were subject to court modification. The real locus of power lay with the court and not with the commission. Furthermore, if this new administrative body wished to revise the maximum rate schedules or minimum standards of service established by statute, it only could advise the legislature.[19]

To emphasize the limits of the legislation of 1905 is to miss its larger significance. This was New York's first successful attempt to regulate gas and electric utilities with a permanent, continuously functioning administrative body. Unlike the rapid transit commissions which were little more than public construction agencies, the gas and electricity commission exercised a measure of control over the practical workings of utility corporations. This represented an important step in a developing progressive answer to the problem of the regulation and expansion of urban economic facilities.

Equally as significant and, moreover, indicative of the urban progressive approach to reform was that the recommendations to the legislature and the bills subsequently passed were framed out of the mass of factual material compiled by Hughes during the investigation. The case for reform was based not on suspicion, righteous indignation, ideology, or social philosophy, but on undeniable evidence. Hughes' conviction that utility rates were too high, corporate profits excessive, and utility-government relations unhealthy was grounded in fact. By an incessant questioning of witnesses and an exhaustive examination of the books and records of utility companies--a vigorous research procedure--Hughes demonstrated his professional skill and obtained the objective data needed to give substance to what before had only been reform rhetoric. Furthermore, Hughes developed a sensitivity to and understanding of utility problems not attainable by attending even scores of meetings at the Social Reform Club or at any other civic organization.

No better recognition of Hughes' contribution to reform could have been tendered him than the request a few months after the gas investigation to serve as counsel for a legislative inquiry into New York's life insurance companies. No better choice could have been made, the World perceptively editorialized, than "the patient and persistent prober of the gas trust abuses. . . . Mr. Hughes would get the facts."[20] That he did and again prepared a detailed report, made recommendations, and helped draft legislation that went a long way toward eliminating abuses in the insurance industry.[21]

Hughes emerged from the gas and insurance investigations "as a public figure of outstanding prominence."[22] He projected on the state and national level the same reputation as a brilliant and honest lawyer that had won him acclaim at the New York bar. Also, in serving as counsel for the legislative committees he first experienced the satisfaction of public service and gained a thorough knowledge of the seamy side of urban business and politics. His insurance investigation exposed the sordid connection between the state Republican party and business and the dishonest practices of several Republican bosses. Hughes' findings caught the attention of President Theodore Roosevelt. When faced with the reality of a gubernatorial election in 1906, an aroused press and citizenry, and a tarnished party image, Roosevelt personally dictated to the Republican convention at Saratoga the nomination for governor of the politically clean and hard-hitting Hughes. The President brilliantly blended the needs of practical politics and a desire to extend progressive government to Albany. Tammany Boss Charles F. Murphy also responded to the popular reform agitation and conceded to the nomination by the Democrats of William Randolph Hearst, a political maverick and perennial foe of bossism and big business. After a vigorous and successful campaign, Hughes was elected chief executive of the state of New York.[23]

"There is no panacea in executive or legislative action for all the ills of society which spring from the frailties and defects of the human nature of its members," conceded the new governor in his first inaugural address. "But this furnishes no excuse for complacent inactivity and no reason for the toleration of wrongs made possible by defective or inadequate legislation or by administrative partiality or inefficiency."[24] Behind Hughes' "unusually strong administrative instincts," lay his insight into the realities of a modern industrial and urban state.[25]

He recognized that increased public demands in the areas of welfare, public health, transportation, and living space meant "a constant extension of governmental activity."[26] For Hughes this required "the strictest insistence upon the highest administrative standards."[27]

Hughes transformed rhetoric into deeds. Ignoring the machine politicians of his own party, he replaced incompetent officials and made appointments to state offices on the basis of ability. His long battle to remove "Honest" Otto Kelsey, New York's superintendent of insurance, was the most celebrated. Kelsey had been appointed superintendent after the Armstrong insurance investigation revealed the failure of the state insurance department effectively to supervise the life insurance business. But after a year in office, Kelsey had reformed nothing and had not even familiarized himself with the findings of the investigation. Because Kelsey "conspicuously failed to perform obvious duties," Hughes recommended to the senate that he be removed from office. In the governor's opinion, Kelsey's personal integrity and attention to the minor details of his department did not "compensate for the lack of administrative capacity." That a man was expected to exercise ability as well as honesty in public office seemed an affront to the organization Republicans who quickly rallied behind Kelsey and prevented his removal during Hughes' first year in office.[28]

Actually, the new governor's expectations complemented well the simultaneous drive for administrative efficiency conducted in New York by Commissioner of Accounts John Purroy Mitchel and the Bureau of Municipal Research. When Mitchel, Henry Bruere, and George McAneny sought the dismissal by the governor of incompetent city officials, they not only found a sympathetic chief executive who talked their language, but one who also was engaged in a similar struggle. Like his fellow progressives in the city, Hughes sought good administration not just by encouraging the use of good administrators but by experimenting with techniques of government that would secure a measure of reform on a continuous and largely self-sustaining basis. Public utility regulation, part of Hughes' broad progressive program that included changes in election procedures, better labor laws, court reforms, and other extensions in police and welfare services, was perhaps the most significant reform of this nature.

In the months immediately after his election, Hughes gave much thought to the problem of more adequate and better regulated gas, electric, urban transit, and railroad facilities. To obtain these ends the governor suggested that the legislature "confer proper power upon a subordinate administrative body." Transportation problems were singled out for special treatment by the governor. "All the existing lines, surface, elevated and subway, are overburdened and the people suffer in mind, body and estate," he told the legislators. In addition to new facilities, he explained, there was an "urgent necessity for more strict supervision to secure better service on existing lines."

To centralize regulatory authority over subways, street surface lines, elevated roads, and gas and electric companies, he recommended that the board of rapid transit commissioners, the state board of railroad commissioners, and the state gas and electricity commission be replaced by a single commission. Because this commission would exercise important state power, Hughes suggested a state board composed of men familiar with conditions in the city. However, he recognized the magnitude of New York City's utility problems and asked for two separate commissions, one having jurisdiction over gas, electric, and railroad corporations in the city alone and the other exercising such power in the rest of the state. Thus "the whole question of transportation, and of gas and electric service, in the territory of Greater New York can be dealt with in an intelligent and efficient manner," he boasted.[29]

Hughes' choice of the commission as a reform weapon was no break with the past. The creation of the state commission of gas and electricity had been his idea. Even his brilliant observation that existing piecemeal efforts at utility regulation needed to be replaced by a comprehensive approach was more the extension of an idea than an innovation. In line with progressive precedents elsewhere, Hughes broke with New York's tradition in his plan to shift the locus of regulatory power from the legislature and courts to the new commission. After enumerating a number of specific powers, the governor recommended that the commissions "direct whatever may be necessary or proper to safeguard the public interests and to secure the fulfillment of the public obligations of the corporations under its supervision." In addition, he asked that the commissions be given the right to institute legal proceedings and that penalties be applied to corporations and officers that disobeyed the commissions' orders.[30] Unaccustomed to such executive initiative, the regulars of both parties received the governor's reform proposal "with the air of sullen anger."[31]

Governor Hughes worked closely with Senator Alfred Page, Assemblyman Edwin A. Merritt, William M. Ivins, and other advisors in preparing a bill. They had as aides the preliminary investigations of the Wisconsin legislature, the experience under the federal interstate commerce act and past New York laws. The Page-Merritt bill, named after its sponsors, created two five-member commissions, one for New York City and one for the rest of the state, each having jurisdiction over all public service corporations in its area except telephone and telegraph; the rapid transit board, the state railroad commission, and the gas and electricity commission were abolished and all their powers transferred to the appropriate new commission. As outlined in Hughes' message, the commissions were granted authority to investigate complaints, fix rates and services on their own initiative, ensure adequate service, establish uniform accounting systems, supervise stock issues, mergers, and consolidations, and subpoena witnesses, and otherwise enforce their orders. The commission for New York City, in addition to its regulatory powers, also was given the task of planning, building, and, if need be, operating elevated or subway lines.[32]

Introduced in March, the public service commission bill

immediately created controversy. Several groups, most importantly the City Club and Merchants' Association, suggested procedural amendments that were later incorporated into the law. The press as well as a host of civic and commercial groups in New York City generally were enthusiastic.[33] Hughes rallied support for the measure by embarking on a two-month speech-making campaign across the state, and in the process compiled a progressive document for effective administrative regulation of public service corporations. The governor attacked the utilities' lawyers' contention that the commission's power to supervise capitalizations, stock issues, operating methods, schedules, rates, and prices was tantamount to confiscation of property.[34] He replied that only an administrative board with "full power to conduct investigations and to make whatever orders in relation to operation that may be necessary" could "secure the performance of public obligation."[35]

To the corporate lawyers and businessmen who argued for a court review of the orders of the commission "both as to facts which may be in dispute and as to the law involved," Governor Hughes lectured that a court was not an administrative body.[36] "It is not the proper function of the courts to fix rates or to make orders as to the facilities," he told an audience in Utica. "This is the function of the Legislature, or of the administrative board which it may create. . . ." For the court to concern itself with any but questions of constitutional or statutory authority invariably meant delay, an unnecessary duplication of proceedings, and "the substitution of the judgment of the court for the action of the commission." Of course, that is precisely what was wrong with earlier experiments in regulation by commissions. This was inefficient administration, Hughes repeatedly contended.

Efficient means of regulation were as important to Hughes as the idea of government supervision itself. Because utility problems were complex and required investigations, the consideration of a multitude of details, and flexible action, "a commission of dignity of force, of ability, representing the best intelligence of the State" was a necessity, he reiterated. The proposed commission "should have such an equipment and such technical assistance as will enable it to deal with the matters before it thoroughly and expertly." In such a scheme the governor, as chief manager of an elaborate bureaucracy staffed with experts, needed extensive powers to hire and fire administrators and oversee laws and their execution. Directly responsible for the conduct of state affairs, the governor, Hughes explained, in turn would be directly accountable to the people.[37]

The fight over the public service commission bill revealed the limits of business support for reform. Members of the Chamber of Commerce, Merchants' Association, and Board of Trade and Transportation, leaders in utility reform a decade before, generally supported the corporate lawyers' arguments for "judicial review as to the justice and reasonableness of orders which tend seriously to impair the income or available capital of corporations." The Chamber of Commerce vehemently opposed as "inconsistent with efficiency and even safety of administration" that part of the law granting the commissioners power over operating and financial matters. "Rigorous publicity is the true, and, in the long run, the most effective cure for evils in these respects, rather than the establishment of minute requirements as to the details of administration," reported the Chamber's committee on transportation.[38] The reform impulse for many of these businessmen still largely was rooted in the mugwump politics of the nineties.

In his fight for passage of the public service commission bill, Hughes often attached to his serious expositions on the need for bureaucratic control of utilities a rhetoric drawn from the mugwump tradition. "I am here under a retainer. I am here retained by the people of the State of New York, to see that justice is done," he boasted in a speech at Elmira. "I stand for the people of the State of New York against extortion, against favoritism, against financial scandal, and against everything that goes to corrupt our politics by interference with the freedom of our Legislature and administration."[39]

While such political evangelism may have elicited specific enough support to force some of the Republican Old Guard to do an about-face and support the utility bill, the timely, if officially unappreciated, intervention by President Roosevelt cajoled the Republican legislators to back the bill as a party measure. The Republicans united to defeat every attempt by the Democrats in the assembly and senate to amend the bill. By May 23 the Page-Merritt bill had passed both chambers of the legislature.[40] A week later it was vetoed by Mayor McClellan. Because New York City had no voice in the selection of the commission for the metropolis, the mayor called it a "piece of blind political legislation."[41] However, the bill was repassed and became law. "In losing the shadow we may have gained the substance of home rule," suggested the New York Times.[42]

"I look upon it as one of the most important pieces of legislation enacted within the last twenty years," former Mayor Seth Low wrote Hughes. "I hope and believe that it will afford the basis for a wiser handling in the future than in the past of the great interests coming under the control of the two commissions."[43] For New York City's progressives, this hope seemed realized. Most important, wrote Henry Bruere a year later, the commission "has adopted the principle of basing its actions not on guesses or expediency, but on facts secured through scientific inquiry."[44] That was the essence of Hughes' triumph. Progressives hoped that this "method of intelligence" would be emulated in other departments of public business.

NOTES

1. Edward A. Fitzpatrick (ed.), Experts in City Government (New York, 1919), p. 31; for a clear presentation of a functional argument for regulation of public utilities, see King (ed.), Regulation of Municipal Utilities, Chapter I.

2. Quoted in New York Times, May 4, 1907.

3. Henry Bruere, "Public Utilities Regulation in New York," Annals, XXXI (May, 1908), 536.

4. William Dunton Kerr, "Qualifications Needed for Public Utility Commissioners," Annals, LIII-LIV (May, 1914), 29.

5. "PSC Report, 1907," pp. 46, 465.

6. Ray Stannard Baker, "The Subway 'Deal,'" McClure's Magazine, XXIV (March, 1905), 463.

7. Mosher, "Public Utilities," p. 236.

8. Wright, "Development of Transit Control," p. 22.

9. Ibid., pp. 22-23.

10. Merchants' Association of New York, Passenger Transportation Service in the City of New York (New York, 1903), pp. v-ix; Merchants' Association, Yearbook, 1905, p. 45, 1907, p. 17.

11. BTT Minutes, October 28, 1903, November 28, 1906; Report of Julius Henry Cohen, Chairman of the Committee on Legislation, to the City Committee of the Citizens' Union, February 23, 1904, pp. 4-5, Box C-1, CU Coll.; New York Journal of Commerce, March 17, 1964, Scrapbook No. 21, 1903-1904, People's Institute Collection.

12. Minutes of Special Meeting of the Executive Committee of the New York Board of Trade and Transportation, August 31, 1905, p. 4.

13. Merlo J. Pusey, Charles Evans Hughes (2 vols.; New York, 1951), I, 74-82, 105-117, 132-133; Dexter Perkins, Charles

Evans Hughes and American Democratic Statesmanship (Boston, 1956), pp. xv-xvii.

14. Quoted in Pusey, Hughes, I, 134.

15. Ibid., pp. 133-138; Henry C. Beerits, "The Gas and Insurance Investigations," pp. 3-7, Box 165, Hughes Papers; Burton J. Hendrick, "Governor Hughes," McClure's Magazine, XXX (March, 1908), 530-531.

16. New York Times, April 22, 1905.

17. Annual Report of the City Club of New York, 1905, pp. 56-57; New York World, March 31, April 29, 30, May 5, 1905; Pusey, Hughes, I, 138-139.

18. Quoted in Mosher, "Public Utilities," p. 238.

19. "First Annual Report of the Commission of Gas and Electricity of the State of New York," New York Senate Documents, No. 15, VIII (129 Sess.), 1906, pp. 64-73.

20. Quoted in Pusey, Hughes, I, 141.

21. Ibid., pp. 140-168.

22. William A. Prendergast, "Reminiscences," CUOHC, II, 226.

23. Robert F. Wesser, "Theodore Roosevelt: Reform and Reorganization of the Republican Party in New York, 1901-1906," New York History, XLVI (July, 1965), 230-249.

24. Public Papers of Charles Evans Hughes, 1907 (Albany, 1908), p. 7.

25. Perkins, Hughes, p. xxi.

26. Hughes Public Papers, 1907, p. 8.

27. Ibid.

28. Ibid., pp. 8, 245-255.

29. Ibid., pp. 29-35; Henry C. Beerits, "First Term as Governor," pp. 23-24, Memorandum, Hughes Papers.

30. Hughes Public Papers, 1907, pp. 30-33.

31. New York Times, January 3, 1907.

32. "PSC Report, 1907," pp. 7-9; Veiller, "Reminiscences," I, 149-150; "Public Service Commission," pp. 1-5, Box 166, Hughes Papers; New York Times, March 6, 1907; Wesser, "Hughes and Urban Sources," pp. 380-381.

33. New York World, March 6, April 9, 22, 29, 1907; Report of the Committee on Legislation of the Citizens' Union, 1907, p. 4; George McAneny to Charles E. Hughes, April 20, 1907, "Memorandum in re amendments to the Public Service Commission's Bill, Proposed by the City Club" [n.d.], p. 1, McAneny Papers; Minutes of the Board of Trustees of the Chamber of Commerce of the State of New York, March 20, April 24, 1907; Merchants' Association, Bulletin, No. 34 (May 3, 1907), p. 6, No. 37 (July 17, 1907), p. 2; Minutes of Special Meeting of the New York Board of Trade and Transportation, April 8, 1907, pp. 1-3.

34. Beerits, "First Term as Governor," pp. 28, 33; New York Times, March 28, April 4, 5, 1907; Wesser, "Hughes and Urban Sources," pp. 382-383; Wright, "Development of Transit Control," pp. 33-34; CC Annual Report, 1906-1907, pp. 120-124.

35. New York Times, April 2, 1907; "Speech of Governor Hughes at banquet of the Utica Chamber of Commerce," April 1, 1907, Box 181E, Hughes Papers.

36. New York Times, March 28, 1907.

37. Ibid., April 2, 1907.

38. CC Annual Report, 1906-1907, pp. xx, 120-123; Minutes of Special Meeting of the New York Board of Trade and Transportation, April 8, 1907, p. 3.

39. Quoted in Pusey, Hughes, I, 205-206.

40. Wesser, "Hughes and Urban Sources," pp. 383-385; New York Times, May 7, 9, 10, 14, 15, 16, 23, 1907; New York World, May 15, 16, 23, 29, 1907.

41. Quoted in New York Times, May 29, 1907.

42. New York Times (editorial), May 29, 1907.

43. Seth Low to Hughes, May 23, 1907, Hughes Papers.

44. Bruere, "Public Utilities Regulation," p. 17.

PART III

THE TRIUMPH OF PROGRESSIVISM

Progressives concerned with improving health and housing, streamlining the government, and regulating utilities shared a common faith in administered progress. By involving government continuously in solving pressing urban problems, they enlarged New York's bureaucracy, created new opportunities for specialists, and concentrated a measure of political power in administration. In so doing they helped modernize New York City's government.

Chapter 8

THE POLITICS OF RESEARCH

Mayor Seth Low's administration "conclusively disclosed the insufficiency of reform by the ballot," declared Robert Fulton Cutting, one of Low's chief supporters. "The man without the machinery, animated by the noblest ambitions, is compelled to pursue them with an antiquated, rusty, unreliable mechanism that paralyzes progress."[1] The Bureau of Municipal Research, which Cutting helped organize, was convinced that its experts, if given the chance, could devise "an efficient technique of municipal administration."[2] Under Democratic Mayor George B. McClellan, Low's successor, bureau progressives got the chance to work with a number of city officials. By the mayoral election of 1909 research-oriented reformers even were elected or appointed to several positions in government. In 1913 John Purroy Mitchel, a progressive politician closely associated with the Bureau of Municipal Research, moved into the mayor's chair. Mitchel's election culminated progressive efforts begun around 1906 to substitute research for reform, process for panacea, and administration for men and measures. In the process these ambitious reformers at least partially restructured and modernized New York City's government.

Efficiency and business methods in government were the watchwords of research-oriented progressives. However, such slogans referred to procedures and not policies. Too many progressives failed to distinguish means from ends and never made clear whether better administration was a necessary precondition for social betterment or merely a penny-pinching affair. This failure limited their effectiveness, tarnished their public image, and shortened their occupation of city hall.

In 1905 Mayor McClellan was renominated by Tammany Hall. Again the Citizens' Union hoped to rally the Tiger's foes behind a fusion ticket but was unable to ignite much support behind its favorite for mayor, District Attorney William Travers Jerome. When deserted by the Republicans, who nominated William M. Ivins, a lawyer and former city chamberlain, the Citizens' Union abandoned the fusion effort and only worked to re-elect Jerome district attorney. The Union never again nominated its own mayoral ticket. Those fusionists who favored the municipal ownership of public utilities joined the camp of newspaper publisher William Randolph Hearst, leader of the newly formed Municipal Ownership League and the only candidate unequivocally pledged to public ownership of major utilities. Hearst received strong support from lower-middle-class voters, traditionally loyal to Tammany, and divided the independent vote with Ivins. Scores of Republicans, scared by Hearst's radicalism and popular appeal and sure Ivins was a loser, voted for McClellan as the lesser of two evils. Even at that, the mayor won the three-cornered race by only 4,128 votes and never lived down Hearst's charge of a stolen election.[3]

In less than a year, McClellan was engaged in a struggle to overthrow Charles F. Murphy's leadership of Tammany. This opened the way for the newly organized Bureau of City Betterment to "apply the process of research, and disinterestedness of research, and the techniques of business management, to public affairs and civic problems."[4] Director Henry Bruere headed an investigation of President John F. Ahearn's administration of the Borough of Manhattan. "How Manhattan Is Governed" accused the borough president of incompetence, inefficiency, and waste. Significantly, little was said about Ahearn personally. Quick to call the bureau's charges lies, Ahearn, a leading Tammany official since 1882, indignantly asked McClellan for an investigation of his office by the commissioners of accounts. The mayor, unsure of Ahearn's loyalty, agreed. He ordered William B. Ellison, the city's corporation counsel, to assign a member of his staff to the investigation; Ellison appointed as a special counsel John Purroy Mitchel, a young lawyer and son of personal friends.[5]

Mitchel, "a brilliant, extremely likeable, and handsome" young man, was glad for the chance to "utilize his legal knowledge and gain added legal experience."[6] Immediately he became the driving force behind the investigation and when after a few months George Von Skal, one of the commissioners of accounts, resigned, Mitchel was given his position. The young lawyer quickly established close ties with the Bureau of Municipal Research and shortly after his appointment invited the bureau to study the office of commissioners of accounts to find ways of making it more efficient and better able to examine city departments. More than anxious to establish working relations with city officials, the bureau seized this chance to promote municipal research. For the office to fulfill its potential as an audit and investigating department, the bureau recommended the installation of cost and efficiency records, time sheets, a system of accounts and improved audit methods, an experienced and technically trained staff, and the addition of an engineering bureau to enable the commissioners to deal with technical questions in the examination of city departments. Impressed with the report, McClellan authorized Mitchel to carry out all of its recommendations.[7] "The service of the bureau in purely municipal work marks a new departure in city government,--the active co-operation of the public with the city administration," declared the mayor at the close of his administration.[8]

In the meantime, Mitchel examined the bureau's charges against Ahearn and conferred regularly with Bruere and other staff members of the new Bureau of Municipal Research as well as with representatives of the Merchants' Association which recently had investigated street pavements in Manhattan. Expert accountants, engineers, and architects were hired to examine the accounts and records of the borough president's office and to investigate street pavements. The testimony of these experts at the public hearings, which lasted from April through July, 1907, fully substantiated the charges in Bruere's original study. When the hearings ended, Mitchel, with Bruere's help, prepared a report which denounced Ahearn's administration for its "inefficiency, neglect, waste and corruption." More to the point, Ahearn was charged, among other things, with not keeping street pavements in repair, buying supplies at exorbitant prices, and padding payrolls.[9]

George McAneny and Henry C. Wright, both officers of the City Club, officially presented to Governor Charles E. Hughes the charges against Ahearn with a petition for his removal. After reviewing the evidence and cross-examining witnesses, the governor on December 9, 1907, removed Ahearn from office. "If there has been maladministration in matters seriously affecting the public welfare and gross breach of official obligation," Hughes declared, "the duty of removal is clear, albeit there is no proof of peculation or personal dishonesty."[10] The Bureau of Municipal Research claimed Ahearn was perhaps the first American official to be removed from office because of incompetence.[11]

The commissioners of accounts then investigated a series of charges filed by the Civic League against Louis Haffen, borough president of the Bronx since 1898 and a politician high in the councils of Tammany. Citing Haffen with "misconduct, incompetency, neglect, waste and violation of law in the administration of his office," the commissioners secured his removal from office by Governor Hughes.[12] "It is not enough for a public official to keep outside the criminal law in performing his duties. He is under obligation to keep his work and subordinates up to a certain standard of efficiency. If he fails in that, he is accountable to the head of the government," declared the jubilant Mitchel.[13]

Mitchel represented one of the earliest links between the nonpolitical municipal research movement and the real world of city politics. "The real usefulness of our office," he declared at a budget exhibit sponsored by the Bureau of Municipal Research in 1908, "will lie not so much in detecting the individual instance of malfeasance on the part of any public official, but will lie in detecting the inefficient method, wasteful expenditure, the spot where waste may and can occur, the method which permits graft, and in pointing out how those defects can be cured, the waste stopped," and more subways, schools, and hospitals built.[14] Allen, Bruere, or Cleveland would not have said it differently, nor for that matter any better. Mitchel's drive, combined with the administrative reorganization suggested by the bureau, enabled him to turn "an insignificant city department into a valuable arm of municipal administration."[15] It also launched him on a political career that in less than a decade would put him in the mayor's chair.

During McClellan's administration the Bureau of Municipal Research also was able to introduce "the techniques of business management" in government. In 1905 Herman A. Metz was elected comptroller on McClellan's ticket. Taking seriously his campaign slogan "A Business Man for a Business Office," he sought to protect the city's treasury from plunder. However, after two years as chief fiscal officer of the city and important member of the Board of Estimate and Apportionment, Metz confessed in a speech before the City Club that without an adequate municipal accounting system he could not manage the mass of business details and therefore could do little to protect the city. When Metz took over the city's finance department there were no records or other means to verify money claims against the city for services or goods, determine the honesty of employees handling city money and supplies, and figure the cost of municipal services and activities. Millions of dollars in liabilities had not been entered in the department's books and no record kept of properties owned by the city. Also, departmental accounts had never been reconciled with what records existed in the comptroller's office.[16] "Like Topsy, the business procedure had 'just growed,'" lamented Frederick A. Cleveland, "and there were in it as many kinks as in Topsy's unkempt hair."[17]

Even the Board of Estimate and Apportionment, which

prepared the annual tax budget and on which the comptroller had three votes, had never correlated "in any specific way the quantities of work to be done with the authorizations of funds by means of which the work was to be carried on." It simply divided the total amount among the different departments and attached only general descriptive titles to the lump sums.[18] In other words, what the city charter called a budget was nothing more than a list of "drawing accounts" or allowances for offices and permitted no central control over spending.[19]

The move to give the Board of Estimate control over fiscal policy was begun in 1906 by the Bureau of Municipal Research. While helping Dr. Hermann Biggs and the health department secure larger appropriations from the board, the bureau staff, with help from an accountant from Haskins and Sells, a leading public accounting firm in New York, analyzed the health department's financial records. Its report, Making a Municipal Budget, criticized the lack of correlation between the Board of Estimate's lump-sum appropriation titles and the actual lines of work of the health department. Although appropriations were grouped under eleven titles, the bureau discovered thirty-four distinct functions performed by the health officials and recommended the appropriation for the department be classified along functional lines. Thus money allotted by the Board of Estimate for tuberculosis prevention and treatment, child care, and milk supply could only be spent for those purposes. The bureau suggested that such a classification would allow the board to exercise central control over expenditures.

Too late to recast the health department's requests for money for 1907, the bureau prepared a classification of the department's current salaries and expenses that accounted for every employee and charged his salary to the division in which he worked. Bruere presented this limited analysis to the Board of Estimate and Apportionment for guidance in making its budget for 1907.[20] Mayor McClellan and Comptroller Metz were impressed as was the Board of Aldermen, which requested that in preparing the budget for 1907 the Board of Estimate not follow "the usual course of making general and unitemized allowance to the general accounts of the various departments." Instead, it wanted "specific appropriations, indicating which estimates are granted and which are denied, in the case of every item in every account, as they appear in the departmental estimate and that the expenditure of each allowance be confined to the purposes as so indicated. . . ."[21]

In 1907 the budgets of only four departments were cast in such a manner as to show the purpose of each appropriation. For 1908, however, the Board of Estimate urged that wherever possible the budget contain a specific item for each class of expenditures. The new bureau of municipal investigation and statistics, the first in "the creation of highly specialized staff agencies," and the Bureau of Municipal Research shared the job of preparing budget classifications and standard forms to be used by all departments in preparing estimates of expenditures.[22]

An improved and uniform municipal accounting system was a necessary concomitant to better budget making and central control over expenditures. Many bankers and investors likewise were concerned about New York's chaotic accounting procedures and urged Mayor McClellan to appoint a group of bankers to study New York's accounting methods. Instead, at the suggestion of the Bureau of Municipal Research and interested members of the Merchants' Association, Comptroller Metz appointed a local "committee on the revision of accounts and methods."

In the meantime, the Bureau of Municipal Research had been studying the finance department's procedures. In a report to the comptroller, it suggested the department be completely reorganized because the comptroller had too many duties and an inadequate staff. With the report as evidence, Metz obtained permission from the state and municipal civil service commissions to appoint to the Department of Finance a staff of twelve expert accountants. Working with the new committee on revision of methods, these accountants devised for New York a system of central accounting control. In 1908 the new system was tentatively installed in four departments, and the next year installation was begun throughout the city and in the Department of Finance.[23]

Metz returned to private business at the end of McClellan's administration in 1909. Under his comptrollership a measure of honesty in government had begun to be institutionalized in the form of a budget and accounting system. Like Mitchel, Metz served as an important intermediary between the independent nonpartisan research movement and the official world of government and politics. His and Mitchel's cooperation with the Bureau of Municipal Research enabled it to function as a staff agency to the city. In fact, when Metz left office he established the "National Fund for Promoting Efficient Municipal Accounting and Reporting" which allowed the Bureau of Municipal Research to make available to other American cities its experience with modern municipal accounting and business methods. The principles behind New York's new accounting system were explained first in a series of "Short Talks on Municipal Accounting and Reporting" and in 1913 published as a text, A Handbook of Municipal Accounting.[24]

The progressives' concern with research and techniques of managing New York's government inevitably affected the mayoral election of 1909. Beneath the outward trappings of an old-style anti-Tammany crusade were new emphases in New York City's politics largely set in motion by the Bureau of Municipal Research and its allies in McClellan's administration. The perceptive American Review of Reviews commented during the campaign that the work of the bureau had made an issue out of practical business methods in the spending of city revenues.[25] The very year before the election, the bureau had sponsored New York's first budget exhibit which, like Veiller's earlier tenement house exhibit, attracted thousands of New Yorkers including the governor, who remarked that it was "an important advance over the old methods of stirring up agitation without really being of assistance." "The exhibit has helped greatly to show taxpayers the superiority of fact over rhetoric,--to teach them to watch the spending rather than to talk about spenders, to demand business methods wherever the city spends or receives money, to judge taxes by their returns, not their size," wrote William H. Allen who was largely responsible for the exhibit's success.[26]

However exaggerated, Allen's claims at least were partially true. Even if many of New York's poorer citizens understood little of Comptroller Metz's accounting system or Mitchel's reorganization of the commissioners of accounts office, they knew Ahearn and Haffen had misused public funds. Tammany's only defense was to call the new research organization the "Bureau of Municipal Besmirch." In 1909 Charles F. Murphy, Tammany's astute leader, plagued by uncovered corruption, dismissed officeholders, and a group of dissident Democrats who only a few years before had sought to remove him as Tammany's chieftain, needed a politically respectable candidate behind whom both factions of the party could unite. In the past, the quiet boss had been a master of just such arrangements, often to the dismay of his reform opponents.

Conveniently, William J. Gaynor, a justice of the state's supreme court for fourteen years, was available. One of Brooklyn's successful lawyers during the 1880s, Gaynor first became involved in politics as the attorney for William Ziegler, a reformer-millionaire, who sued the city of Brooklyn for alleged public corruption. With his political appetite whetted, Gaynor in 1893 aspired to be Brooklyn's mayor, but instead was chosen by the Republicans for the state supreme court. During the campaign Gaynor launched an attack that brought the downfall of John J. McKane, notorious boss of the district that included Coney Island, and extended his own reputation to Manhattan and upstate. As a judge with an acid tongue, Gaynor kept his name before the public and often was mentioned as a candidate for elective office. Although a Democrat in national politics, Gaynor steered clear of political labels on the local level. At the age of sixty-one, respected, regarded as a champion of the public interest, and generally mistrusted by politicians, he was tagged by the unpredictable Murphy as the man to carry the Tiger's banner in 1909.[27]

From a distance Murphy's opponents appeared much like the old reform crowd. A large group of lawyers, businessmen, bankers, and labor leaders formed a Committee of One Hundred which held mass rallies, negotiated with civic organizations including the Citizens' Union, and sought to unite Tammany's enemies behind a single slate of candidates. Before these gentlemen, several of whom were veterans of the campaigns of the nineties, could agree on suitable nominees, however, William Randolph Hearst ordered his spokesmen to abandon the muddled fusion effort and the Republicans, also cooperating with the Committee of One Hundred, held their nominating convention. Republican leaders, aware of Gaynor's support among independents and of their party's minority status in the city, conferred with spokesmen from the Citizens' Union and Committee of One Hundred before nominating Otto T. Bannard, well-known businessman, banker, and social and charity worker, for mayor; John Purroy Mitchel for president of the Board of Aldermen; William A. Prendergast, former secretary of the National Association of Credit Men and present registrar of Kings County, for comptroller; and George McAneny, independent Democrat, for president of the Borough of Manhattan. These candidates were accepted by the Committee of One Hundred and a Republican-Fusion alliance cemented.[28]

Significantly, two of the candidates, George McAneny and John Purroy Mitchel, were closely associated with the research movement. Mitchel, campaigning for the presidency of the Board of Aldermen, promised to continue the improvements begun by the McClellan administration. As a member of the Board of Estimate and Apportionment, he told his audiences, he would be strategically located to give New Yorkers a business administration. To get more subways, schools, housing, and other welfare improvements, he reiterated, the city needed to employ business methods.[29]

In another speech Mitchel outlined an efficient program for New York's government that his friends at the Bureau of Municipal Research could hardly have improved.

> 1st: A knowledge of the facts, the full facts, all the facts, as a prerequisite to the appropriation of funds by the Board of Estimate. 2nd: An informing, accurate and controlling system of accounts focused into the Department of Finance. 3rd: The standardization of contract specifications. 4th: Honest, expert and prompt inspection of contracts. 5th: Centralized purchasing of municipal supplies. 6th: Accurate cost records in all departments for contract as well as payroll work, whereby results may be compared and judged. 7th: Examination and analysis of performances and expenditures conducted by vigilant and active commissioners of accounts. 8th: Cohesive action by an earnest and harmonious Board of Estimate in applying this program from January 1, 1910 to December 31st, 1913.[30]

Little did Mitchel realize then that in 1913 he would carry the torch of the "efficiency and business in government" campaign and this program would in essence be the platform for his own mayoralty campaign. Even New York's better citizens emphasized the business side of government and politics: "The need of securing an administration of the affairs of the City of New York based upon business principles is plainly apparent," stated the call for the mass meeting that selected the Committee of One Hundred.[31]

In "'The Business of Citizenship' in New York City," an article written shortly before election day, 1909, Allen tried to capture the change occurring in New York's politics. "It is because of New York's achievements that she sees her defects so clearly," he wrote.

> We discern the promised land; but not having been there, our path ahead is not well cleared, signboards are not yet up, landmarks are missing. Leadership which was effective with an uninformed, stampedable public is itself a stranger in this new pioneer world. The "psychological moment" school of political pilots were adept in handling "now you see it, now you don't see it" issues on waves of unrest; but they have not yet gotten their land legs.
>
> There was a time when New York did not know what it wanted. This year so many of us know not only what we want done but how we want to get it done that a new kind of leadership is required.
>
> . . . .
>
> Without fully realizing it New York City has set the nation an example by waking up to the fact that what may happen from 1910 to 1914 will depend so largely upon outside contractor, outside political organization, and inside methods of serving the contractor and organization, that it has made the vital issue <u>what these three combined are apt to do</u>. Vaccinating the public by bogus ante-campaign mass meeting of protest against waste has failed to make the city immune to real smallpox-evidence in the midst of the campaign. The public has shown capacity to use information.
>
> This new knowledge is specific as never before with regard to (1) waste, (2) inefficiency, (3) steps recently taken, and (4) steps that must

> yet be taken to reduce and to prevent future waste and inefficiency. While civic bodies have played an important part in bringing this situation about, the fullness of the opportunity is due to the present administration, its improvements and investigations.[32]

Because Bannard had little public appeal and Gaynor was immensely popular, the Republican-Fusion people sought Hearst's support. When Hearst refused to endorse Bannard, the Republicans persuaded him to enter the contest for mayor as the candidate of his newly formed Civic Alliance, the successor to the defunct Independence League, and to endorse the rest of the fusion slate. This political strategy in the main succeeded. Gaynor won the three-cornered mayoralty race, defeating Bannard, his nearest opponent, by over 71,000 votes. But Bannard and Hearst together piled a total vote almost 80,000 larger than the judge's, and this anti-Tammany majority elected the entire fusion slate below mayor except the candidate for president of the Borough of Queens.[33] "I am very well aware that until you consented to enter the field there was little hope for the success of the Fusion candidates," Mitchel wrote Hearst. "The citizens of this city are indebted to you for your public-spirited action which will result in protecting the city from the depredations of Tammany Hall during the ensuing four years."[34]

As the fusionists moved into official positions in New York's government, they carried with them the "municipal research idea." In so doing they continued and expanded the administrative changes begun under McClellan's mayoralty. "Systematizing public business has been necessary to equip city government to do the work already committed to it and to prepare it for the assumption of increased responsibilities," explained Henry Bruere.[35] As outsiders recommending better business methods to city officers, bureau progressives needed to give little thought to specific "responsibilities" or city needs.

When they filled official positions, however, they exposed their rhetoric to the harsh reality of meeting New Yorkers' needs and the test of practical politics.

Mayor Gaynor, who generally cooperated with the fusionists, surprised many with his administrative appointments. Most of the commissioners chosen were "distributed among a group of experts and business men, many of whom had probably never seen the inside of the Fourteenth Street Wigwam."[36] Even the few Tammany men chosen were reputable. Many well-publicized problems occupied the Gaynor administration--new subways, the shooting of the mayor, the garment industry general strike, and anti-vice and crime crusades--but perhaps those closest to the hearts of the reformers related to changes in fiscal administration and economy. Even the important fight over city versus private subway lines, which occupied much of the Board of Estimate's time and split the fusion ranks, related in part to questions of municipal finance.

Eager to continue the war against administrative and financial chaos, fusionists exploited the services of the Bureau of Municipal Research. Comptroller Prendergast recalled that during his eight years in the finance department the bureau "played a prominent part in New York City's affairs."[37] Henry Bruere, who worked closely with Mitchel and McAneny, frequently served as liaison man between the Board of Estimate and the research organization. The fusion members of the Board of Estimate came to its first meeting armed with resolutions prepared by the Bureau of Municipal Research and "delivered to the Board-of-Estimate members' baskets on the morning of the meeting."[38] The board passed nine of these resolutions that further systematized city business and inadvertently set a pattern for much of the board's work over the next four years. The city's newspapers, unused to such a flurry of activity, took special note of the business atmosphere of this first fusion board meeting.[39]

The comptroller's office again was the center for reform in New York's business methods as Prendergast tackled his job "with the vigor and intelligence of an experienced business man."[40] He paid all city bills promptly and stopped the practice of meeting current expenses out of the proceeds of long-term bonds. From a Tammany-controlled legislature he obtained a bill providing for the semi-annual collection of taxes, which lessened the city's need to borrow in anticipation of tax revenues. Another measure, passed in 1910, permitted the issuance of new short-term revenue bills, which, unlike city bonds, were more readily marketed in Europe where interest rates often were lower than in America. The comptroller also obtained permission to market bonds only once a year, thus allowing him to select the most advantageous time and incidentally avoid upsetting the bond markets by too frequent sales.[41]

Most important, Comptroller Prendergast in cooperation with the Board of Estimate completed the accounting system and budget reforms begun under Metz. Until about 1913 the trend in budget-making was toward greater segregation of items; that is, the tax budget was analyzed in great detail according to the object of expenditure. Such accounting control, applied initially to salary and wage appropriations, was extended after 1910 to supplies and materials and then broadened to cover salaries and wages paid out of stock proceeds or other sources. Prendergast organized the committee on the standardization of supplies which, under the direction of the Board of Estimate, worked to standardize supply purchases hopefully to replace hundreds of specifications and unit prices with "system and business methods" and thus establish the basis for a central purchasing plan. Complementing this the Board of Estimate further tightened control over purchasing and contract by attaching to the departmental appropriation various rules governing spending. In the budget of 1913 twenty-four conditions were established.[42]

As 1913 approached, the segregated budget increasingly came under fire from the Bureau of Municipal Research and others. Not efficiency but honesty had been institutionalized, declared Charles Beard. The highly functionalized budget "was prompted not so much by the desire to expedite the performance of public business," wrote Henry Bruere, "as to prevent age-long and conspicuous misuse of public funds which under lax organization and ineffective administration had become characteristic in New York. . . ."[43] Only as a matter of expediency, then, had priority been given to central control over departmental expenditures. As early as 1909 budget reformers realized that such a procedure placed too many limitations on executives and resulted in "a degree of regimentation which restricts and in a measure paralyzes the freedom with which the organization provided in the appropriations may be employed, or the funds for purchases may be utilized." Most important, the highly segregated budget alone could not give financial officers sufficient information to determine current costs of doing work. So, after 1913 reformers worked to prepare a budget that allowed greater administrative discretion and at the same time promoted efficiency.[44]

To carry out a recommendation of the Board of Estimate, Comptroller Prendergast's office in 1910 prepared New York's first corporate stock budget which allowed central accounting and administrative control over monies obtained from the sale of long-term securities and spent for permanent improvements.[45] The Department of Finance itself underwent a "reorganization from top to bottom." Formerly separate jurisdictions, each with its own books, records, staff, and accounting and auditing functions, were centralized and "every step in the process of audit and accounting definitely prescribed, and necessary information recorded in one place for the common use of every officer or interested employee of the department."[46] A system of uniform methods of recording and reporting transactions also was extended throughout the city departments and new accounts were devised and installation begun which showed the actual expense of all classes of services and supplies. Accomplished successfully in 1912 was the reconciliation of the Department of Finance's fifteen thousand accounts with the records in 130 outside departments, bureaus, and agencies. Prendergast was aided by a staff of accountants from the Bureau of Municipal Research, and the committee on finance and currency of the Chamber of Commerce reviewed the results. Thus by 1913 the comptroller's office was able to issue to the Board of Estimate monthly reports on the financial condition of the city, allowing that body more intelligently to make decisions on public policy.[47]

Because of victory in 1909 a few research-oriented progressives had moved from outside New York's government to positions of official leadership. Eagerly they anticipated being joined by others after a more complete victory in 1913. Again their vehicle would be an organized citizens' anti-Tammany crusade that appealed to the public, employed a familiar and traditional rhetoric, and obscured the shift in reform emphasis and activities.

Joseph M. Price, a dress manufacturer interested in civic reform and secretary of the fusion committee in 1909, initiated a movement in December, 1912, that in a few months culminated in the formation of a Citizens' Municipal Committee. To assure a wide hearing and broad appeal among anti-Tammany groups, the 107 members were carefully chosen to represent the five boroughs and all political parties. Although none of the city's civic and business organizations was officially represented, many of the committee members were active in these groups.[48] These well-heeled Gothamites vowed "to secure during the next four years a non-partisan, efficient and progressive government for the City of New York."[49] Of the more than 1,000 contributors to the fusion effort, businessmen, lawyers, and bankers were heavily represented. Andrew Carnegie, William H. Childs, R. Fulton Cutting, Cleveland H. Dodge, William Loeb, Jr., George W. Perkins, John D. Rockefeller, Jacob H. Schiff, and Isaac N. Seligman headed the list.[50]

A committee on candidates held hearings at which a number of business, civic, and labor organizations argued on behalf of one candidate or another. The best the committee could do was narrow the choice to three leading men. The favorite of the good government people, especially in the City Club, Citizens' Union, and the settlements, was George McAneny, independent Democrat and president of the Borough of Manhattan. He had been president of the City Club, trustee of the Bureau of Municipal Research, and chief spokesman for the city in the dual subway contract negotiations. "Mr. McAneny is my candidate for the Mayoralty," wrote Lillian Wald. "He understands our language, has social training and experience, and is the kind of public servant that . . . ought to have recognition from those who believe that the city should get the benefit of trained people. . . ."[51]

Progressive party leaders, labor, and the Hearst crowd supported John Purroy Mitchel, recently appointed collector of the Port of New York by President Woodrow Wilson. The Progressives considered Mitchel "an engaging character, of the required forcefulness and the necessary progressive tendencies," and Hearst liked his strong support for city-owned subways. The Republican choice was the popular District Attorney Charles Whitman, who had gained much publicity for his role in the prosecution of police Lieutenant Charles Becker for the murder of gambler Herman Rosenthal.[52]

Determined to choose a slate before the regular parties held their conventions, the deadlocked committee on candidates presented the three names to the executive committee which put them before a meeting of the entire Municipal Citizens' Committee. After eight ballots, McAneny, consistently in third place behind Mitchel and Whitman, dropped out of the race. On the ninth ballot Mitchel received a majority and was declared the nominee. The committee then nominated McAneny for president of the Board of Aldermen, Prendergast for comptroller, and Whitman for district attorney. After some hesitation, McAneny and Whitman accepted.[53]

Mitchel was a compromise selection, suggests his biographer. Only Mitchel of the three candidates had Hearst's support. Municipal reformers liked his record as commissioner of accounts and president of the Board of Aldermen, while

Progressive party members admired his political independence. Even President Wilson gave Mitchel his blessing.[54] Significantly, two of the three nominees, McAneny and Mitchel, were not politicians of the usual sort, or even gentlemen-reformers. Both were closely associated with the Bureau of Municipal Research and in government were spokesmen for the "municipal research idea." In fact, in Mitchel's drive for the nomination, Henry Bruere was one of his closest advisors and McAneny had worked closely with Bruere in reorganizing the government of the Borough of Manhattan. On the fusion ticket the three top city-wide positions--mayor, comptroller, and president of the Board of Aldermen--were filled with men representing the newer research-politics, men more interested in the process and methods of government than in personality. They had more in common with Biggs, Hughes, Veiller, and Cleveland than with Low, Strong, and the Committee of Seventy.

The Fusion nominees were later officially accepted by the Republican and Progressive parties and several groups of independent Democrats. However, the unpredictable Hearst cabled a message to his Independence League, which had endorsed Mitchel's candidacy, that under no circumstances would he accept McAneny and Prendergast on his party's ticket because they had supposedly abandoned election pledges of 1909 and voted for the dual subway contracts. "I intend to assail all public traitors, all corporation mercenaries, no matter under what banner the scarlet livery of their shameful service may be found," he is alleged to have declared.[55] Mitchel then graciously declined the League's nomination believing it would needlessly split the fusion forces and aid Tammany. When the Independence League's second choice for mayor, James A. Allen, resigned late in the campaign, leaders of the Committee of One Hundred and Seven and the Republican and Progressive parties urged Mitchel to reconsider and allow his name to be on the League's ticket. "You are the recognized champion of liberal municipal policies," wrote Francis W. Bird, county chairman of the national Progressive party, and ought to accept the second offer. Henry L. Stimson believed Mitchel's acceptance would "help and not harm the Fusion campaign." Only after the executive committee of the Citizens' Municipal Committee formally allowed all its candidates to accept any nominations offered them did Mitchel accept Hearst's offer.[56]

Not all of the surprises were on the fusion side. At Delmonico's Restaurant three weeks after Mitchel's nomination, Tammany, pressured by the Brooklyn organization, decided to "unmake the gadfly" and bypassed Gaynor to nominate the politically more reliable Edward E. McCall, former judge and current chairman of the public service commission for New York City. Then to neutralize the police corruption issue, the ever astute Tammany leaders endorsed the candidacy of Charles Whitman for district attorney. Willing to fight back, Gaynor accepted an independent nomination at the hands of his business supporters, led by the banker Jacob H. Schiff.[57] Six days later, on a European cruise, New York's mayor, "a frail old man bundled up in a rug in his deck chair," was found dead by his son.[58] Gaynor's supporters offered the nomination to George McAneny who refused. Leaderless, the majority of Gaynor's men moved over to the fusion camp, thus patching the fissure that could have guaranteed a Tammany victory.[59]

Most of the campaigning in 1913 was couched in the rhetoric of reform politics. When accepting the nomination, John Purroy Mitchel asserted he was leading a struggle "between the forces of decency, honesty and public service, and the forces of public plunder." "The dominant issue in this campaign is a very plain issue," he continued. "Shall the people of this city govern it, or shall Mr. Murphy govern it?" Over and over, with words that could very well have been lifted from Harper's Weekly editorials of a decade before, Mitchel admonished his listeners that Tammany was the chief issue of the campaign. Pressing issues--charter revision, home rule, public utility regulation, police corruption, government reorganization, and the need for additional municipal services such as schools and recreation areas--while discussed, were subordinated to the war on Tammany.[60]

The continuation of the politics of reform into the election of 1913 obscured for contemporaries as well as later historians more fundamental changes in government and politics. It was not just another in a long line of righteous crusades against the Tiger. "No good government campaign along old lines will receive the support of the Progressives this fall. . . . The fight must be on issues far more vital than the old slogan of 'milk-diet' reform, 'we're good and they're bad, so vote for us,'" declared the Progressive party of New York. "We are not going into any 'goo-goo' campaign of the old order of reform movements."[61]

In fact, in the same speeches where Mitchel flayed the Tiger, he also promised the voters a business government. "We mean that we will apply to the conduct of the city's business the same zeal, effectiveness and eagerness to serve, that experience has demonstrated are required in private business to secure a dollar's worth of result for every dollar spent," he told one audience. Behind even the familiar jargon of a business government lay new meaning and content. In the same speech Mitchel then defined the content of such a government as consisting of the administrative and financial changes brought about by the Board of Estimate of which he was a member. It was in this machinery, as he labeled the business reforms the fusion board had developed, that Mitchel placed his hope for further progress. When the young candidate argued that the issue was "Murphy, Tammany Hall, contract government, on the one side, or business government for the service of the people on the other," he meant that a business government was not just businessmen or other well-meaning gentlemen in office, but machinery, methods, process--budgets and budget committees, accounting systems, efficiency commissions--that could be used to achieve a better and more responsive municipal government.[62]

Revealingly, all of the major party platforms, while differing in emphasis or details, occupied much common ground. Tammany dared not promise the voters to undo the administrative and financial innovations modernizing New York City's government. All of the parties promised to continue the trend, the Progressive party presenting the most elaborate blueprint for new bureaus or departments and government powers that would expand greatly New York's bureaucracy and influence the lives of all her citizens.[63]

The changes in New York's business methods put into effect from 1906 to 1913, Bruere explained, were "building up in New York City for the first time in any city in America an efficient technique of municipal administration." "New York's next mayor," he further explained, "despite the inherent difficulties of his position, will have an infinitely better opportunity to succeed than any of his predecessors, for the government which will be placed in the hands of the new mayor on January 1, 1914, will be a completely different institution from the government put into Mr. Gaynor's hands in 1910, or in the hands of Mr. McClellan in 1903."[64] What the new mayor would do with a more streamlined government was not explained. The paucity of serious discussion about New York's needs or the fusion candidate's vision for a better city suggested that perhaps these reformers already had mistaken means for ends, had already let unquestionable improvements blind them to the limits of their efforts. This developing flaw in bureau-type progressivism became consequential after election day.

About three weeks before, Democratic Governor William Sulzer was impeached by the Democratic-controlled legislature. Both the governor's supporters as well as the fusion candidates placed the blame for this unusual action on Murphy's dislike of Sulzer's independence. With the Democrats thus feuding and the anti-Tammany vote united, Mitchel won an overwhelming victory, defeating McCall by 121,000 votes. The fusion forces won control of all the city-wide posts, the Board of Estimate, and, most dramatically, even the Board of Aldermen.[65] Just as Low's campaign and administration culminated the mugwump crusades for good government, Mitchel's culminated progressive efforts to substitute research for reform and administration for men and measures. What in 1905 had been only a municipal research "idea," by 1913 had fully captured the government of the city of New York.

In control of New York's government and fully responsible for its policies, the reformers had a rare opportunity to utilize "the machinery of government in dealing with public problems of community welfare."[66] However, they transformed their particular reform virtues into vices by giving too little thought to the ends of government and the needs of people. These well-meaning reformers failed to perceive that techniques of government--"business methods, efficient accounting and the technical aspects of budget making, etc."--however useful, were not the answers to New Yorkers' needs.[67] For many progressives during Mitchel's administration the shibboleths of research politics--economy, efficiency, fidelity, standardization, systematization, and the like--hardened into immutable principles by which reality was measured.

The main thrust of Mitchel's mayoralty was to improve the machinery of New York City's government. He appointed as heads of departments men with special professional training in their areas of service. Dr. S. S. Goldwater, superintendent of Mt. Sinai Hospital, was appointed health commissioner. Goldwater forced all heads of divisions within the department to give up private medical practice and devote full time to the municipal health service. Medical and sanitary inspection was expanded to include factories, stores, restaurants, and transit facilities. In general, Goldwater brought administrative expertise to a department run by doctors. Commissioner of Accounts Leonard Wallstein organized a division within his office to aid in studying administrative problems. The commissioners of accounts office became the "Mayor's Eye."

Under Commissioner John A. Kingsbury, the charities department improved its procedures and sought to better the conditions of municipally controlled charitable institutions and child-care agencies and remove incompetent supervisors. After an investigation of private religious child-care agencies receiving city money, Kingsbury secured improvements in those facilities. A welcomed measure of honesty and efficiency permeated even the police department under Commissioner Arthur Woods.[68] By improving the administration of departments under his control, Mayor Mitchel contributed to the cause of good government.

Several department heads, the Department of Finance, Chamberlain Bruere, the Bureau of Municipal Research, and various committees, all generally supported by the Chamber of Commerce, Merchants' Association, City Club, and Citizens' Union, either singly or in combinations were responsible for a number of studies and administrative reports. These led to the creation of a central purchasing plan, a central payroll division in the finance department, and a consolidation into one department licensing and inspection functions previously scattered among a number of city bureaus. Because of a study by the commissioners of accounts office, untrained coroners were replaced by physicians. A detailed report analyzing the functions of New York's 103 municipal departments and bureaus led to budget cuts and several administrative reorganizations. The efficiency bureau and the budget and advisory staff of the commissioners of accounts office helped modernize the preparation of the finance department's estimates.[69]

In budget-making and accounting procedures the trend was toward greater administrative control. The Board of Estimate, in addition to establishing a subcommittee of its main committee on the tax budget, created a number of highly specialized, advisory staff agencies--bureau of contract supervision, bureau of standards, committee on education--to study continuously problems of buying and using services and materials. In 1914 the board asked the Bureau of Municipal Research to study New York's budget procedures. The bureau's report discussed the budget more as an instrument for planning, a tool of management, than as a means for legislative control over expenditures.[70]

> There has come about in a remarkably short time a complete alteration of the attitude of the public officer to his executive responsibility. Instead of mere opportunism, evasion and compromise to which the average public officer of a decade ago was driven by political conditions or by lack of method and organization, department heads are now seeking to increase, day by day, the quality of service performed by the organization responsible to them, and to exercise increasingly effective discretion with respect to the work for which they are responsible. Because of the exceptional character of its personnel, its conspicuousness, and the ease with which the public could deal with it, the board of estimate, the financial board of New York City, for years served as a shield between the political rapacity or administrative slovenliness of department heads and the public interest. Now, more and more, department heads are better prepared to exercise intelligence with respect to the administration of the board of estimate, to acquire through personal contact or the advice of advisory staffs.[71]

The bureau suggested abandoning the extreme functionalization of the budget in favor of lump-sum appropriations classified only by object of expenditure, such as wages and salaries. "The purpose of this change," the bureau explained, "is to expand the now restricted discretionary authority of department heads and to make more flexible the use of departmental organization."[72] To enable administrators to compare work performed with a given expenditure and thus enable them to estimate future costs for work planned, the bureau stressed a unit-cost accounting system.[73]

By the end of Mitchel's administration, this trend toward greater executive responsibility and flexibility was only beginning to be accepted. The Board of Estimate especially was wary of any change that might lessen its control over all expenditures.[74] "The budget could be the means whereby a legislature could control an unruly, immoral executive; or it could be the means of freeing a purposeful executive from the confusion of legislative indecision," Barry Karl has perceptively noted.[75] Even Bruere and Allen, who by this time was no longer with the Bureau of Municipal Research, split on this issue. In fact, the controversy over the nature and use of a budget reflected a more basic division among reformers. The whole economy and efficiency drive--the research idea--could be understood either as a penny-pinching affair or as a necessary effort to modernize city government to enable it to respond to the modern needs of its citizens.

"There was too much emphasis on economy and too little attention devoted to the needs of a growing city" during Mitchel's administration, suggests Edwin R. Lewinson, the mayor's biographer.[76] In 1914 when the Mitchel administration, as part of a bargain with city banks for help in floating a bond issue to raise eighty million dollars of gold for the city to repay foreign debts, agreed to finance city improvements of a non-income producing nature from annual tax revenues only, several fusion members on the Board of Estimate sought to make this "pay-as-you-go" plan a permanent part of New York's fiscal policy. The reformers hoped such a plan would decrease the amount of city borrowing, save interest, reduce the budget, and perhaps allow a tax cut. That needed city improvements would be curtailed did not enter their calculations.[77]

In a study of Mitchel's career and mayoralty, Lewinson amply demonstrates how welfare and education services too often were sacrificed on the altar of efficiency and economy. The accomplishments of Commissioner John A. Kingsbury in the Department of Public Charities were marred by his excessive concern with economy. Hundreds of feeble-minded and other dependent children in charitable institutions were sent home to their families to save the city money. Mayor Mitchel considered this a signal achievement. In fact, Kingsbury made it clear that he favored private over public charity.

Yearly the Board of Estimate appropriated less money than the school board demanded and deemed necessary to provide the city's children an adequate education. When in 1915 the board of education requested additional money for summer school teachers, Comptroller Prendergast suggested the board get volunteers. Adding insult to injury, the tight-fisted comptroller even accused New York's teachers of laziness. The Board of Estimate also refused to grant sufficient funds above normal budget requirements to keep all the city's evening high schools operating and to have constructed needed school buildings. Such stinginess, even when rationalized as economy, resulted in much unfavorable publicity for the Board of Estimate and the Mitchel administration.

New York's school teachers were alienated further when the mayor sought to gain from the state control over teachers' salaries. While such a move was consistent with the Board of Estimate's efforts to centralize control over all city expenditures, the teachers feared it would be the first step toward lower salaries. Mitchel's fight to reduce the size of the board of education from forty-six to nine members also stirred opposition among teachers as well as parents and the board of education. They believed the mayor would be able to dominate a smaller board and thus control the city's school system, suggests Lewinson.[78] Of course, that is just what the progressives wanted. The bill reducing the size of the board of education was "designed to secure better control and a more business like conduct of the affairs of that Board," Mitchel wrote.[79] Prendergast believed "that the board need not consist of more than five members,--a small, strong, working board of trustees, filled with the idea of their trusteeship and unswayed by the political machinations of those within or without the educational system."[80] Not until 1917 did the state legislature pass a bill reducing the size of the boards of education for all cities in the state.

In 1914 the mayor, the president of the board of education, and other members of the administration visited a number of schools in the Midwest to study experiments in vocational and industrial education. They were intrigued with a system of education worked out by William A. Wirt, superintendent of schools in Gary, Indiana. Elementary school children studied traditional academic subjects for only half of the day. The other half was spent in various vocational, art, and music sessions held in the school auditorium, library, or playground or even in neighborhood settlement, factory, or church facilities. The Gary plan essentially was a "work-study-play" scheme for enriching the curriculum. However, the mayor was most impressed by the possibility for double sessions in each school.[81] "The Gary Plan was meant to improve the quality of education, but members of the Board of Estimate were primarily interested in its economy aspects," writes Lewinson.[82] Comptroller Prendergast made this very evident. "The Gary plan not only proposes vast economies in expenditures for school buildings, but it also points to a substantial reduction in the cost of teaching service."[83]

After experimenting with the Gary plan in a few New York schools in 1914, the Board of Estimate, excited about the money-saving possibilities, unilaterally voted to use the double session system in all of the city's congested schools. Because the educational soundness of the plan had not been determined, the Board of Estimate ran into a storm of opposition from educators, the board of education, school administrators--Prendergast called them "the 'mossbacks' in the educational system"--and parent and teacher organizations. When Tammany Hall again ran the city in 1918 the new Board of Estimate ended the experiment. "The failure of the Gary Plan was due to the fact that the Board of Estimate considered it not so much an educational plan as a device for saving money," concludes Lewinson.[84] It clearly revealed a weakness inherent in bureau-type progressivism and further limited the reformers' political appeal.

Efficiency pushed too far became a tyranny that inhibited the meeting of needs. Too much of what these good-government progressives created were merely tools--"structure, process, technique, and methodology of administration"--that could be used for many purposes.[85] For many businessmen and property-owners, the promise of modernized government was not better services nor new responsibilities but smaller budgets and lower taxes. Even Mitchel's and Prendergast's speeches spoke too often of tax relief and reduced spending to encourage those envisioning a "new city government" that would be the accepted common agency for "getting things done." Budget and accounting systems, purchasing plans, and new bureaus of efficiency prevented graft and institutionalized a measure of honesty in government but could not be eaten, used, or enjoyed by the public. After all, New York's stockholders, to use the jargon of the research crowd, wanted dividends in addition to annual reports.

Progressives expanded the city's bureaucracy, not its services. Every new department, bureau, or study group created to measure, count, or weigh one problem or another meant new opportunities and jobs for trained specialists of all types. Too often, however, the fruit of their work was scores of unskilled workers without jobs. "New York's Government by Experts," as Albert Shaw entitled an article in 1914, became too much a government for experts as well.[86] Enamored of perfecting the city as a machine, these progressives forgot people and lost sight of the ends of government. Like the mugwumps, they too could have learned something from Plunkitt. "You must study human nature," he had said, and "to learn real human nature you have to go among the people, see them and be seen." The city was made up of human beings, not organizational charts and statistical lists. Also, as Plunkitt reminded his listeners, the poor had more friends in their neighborhoods than the rich had in theirs.[87]

John Purroy Mitchel and the fusionists, who had won a smashing victory over Tammany in 1913, were soundly defeated for re-election in 1917. In 1913 Mitchel had received 57 percent of the total vote; in 1917 he received only 23 percent. Tammany's candidate for mayor, former Kings County Judge John F. Hylan, received 313,956 votes to Mitchel's 155,497. Morris Hillquit, the Socialist candidate, received only approximately 10,000 votes less than Mitchel; William M. Bennett, the Republican candidate, trailed behind with 56,438 votes.[88] Why had New Yorkers repudiated Mayor Mitchel? "The failure of Mitchel and the men around him to carry the public with them constituted the chief failure of the Mayor, his commissioners, and the managers of his campaign," writes Lewinson. "He saw economy, efficiency, and social reforms as ends in themselves rather than as means leading to the end of human comfort, welfare, and betterment."[89]

This inherent flaw in Mitchel-bureau progressivism, perhaps serious enough to deny him re-election in any case, was obscured by administration policies that alienated important voting blocs in New York and by fortuitous events during the campaign. The school struggles, in part stemming from Mitchel's excessive concern with economy, had alienated countless parents who felt their children were being neglected and teachers and educators who resented having their services measured by a commercial yardstick. While investigating the performance of religious child-care agencies receiving city money, Mitchel and Commissioner Kingsbury alienated the Catholic church which resented the whole affair. The mayor's image had been tarnished further when the news leaked that the administration had tapped the telephone wires of a Catholic priest and others involved in the charities dispute.[90] "If the same thing had been done by Mayor McClellan," William H. Allen wrote, "there would have been a riot in New York."[91]

Excessive unemployment and high food costs stemming from World War I further clouded reform's future. Poor-people's marches, riots, and general dissatisfaction greeted all of the mayor's efforts to relieve the hungry and jobless. Mayor Mitchel's choice of Elbert H. Gary of United States Steel to head a committee on unemployment and George Perkins, the financier and leader of the Progressive party, to chair a committee on high food prices was made in good faith, but they were hardly the men to inspire confidence among the poor who needed relief. Their patronizing attitude and the mayor's limited interest in the work of the committees negated any public appreciation Mitchel could have expected.

Mitchel was an early advocate of national military preparedness, a position championed by Republicans and the wealthier segments of New York's population but largely ignored or even rejected by the majority of New York's citizens. The mayor passionately preached his convictions, alienating for a time President Woodrow Wilson, argued for universal military training, and even joined an army camp for a month's military exercises. As the issue of national preparedness intensified, so did Mitchel's patriotism.[92] "The time is approaching when the people of America will be divided into two parties--Americans and traitors," he told an audience in 1917.[93] In a conflict with State Senator Robert F. Wagner over the proper procedure for the city to follow in making available to the federal government land at Rockaway point for fortifications, the mayor branded Wagner a tool of the Germans. In fact, patriotism was Mitchel's major campaign theme and it backfired as numerous Socialists who opposed the war and German- and Irish-Americans who opposed America's seemingly pro-British stance deserted the fusion cause. Mitchel's effort to label John Hylan as pro-German and therefore un-American aroused criticism and much sympathy for Tammany's mayoral candidate. In turn, the Tiger managed to smear Mitchel as a captive of the rich. It was a question of the people versus money, Tammany argued.[94]

Mitchel's supporters ran a poor campaign. The fusion committee of 250, led by men of wealth, had little political experience and even less knowledge of New Yorkers' needs. They operated under the illusion that the Mitchel administration had "executed an unprecedented program of government for all the people."[95] So sure were they of Mitchel's re-election that they convinced the mayor he need not take seriously former state legislator William M. Bennett's challenge in the Republican primary. However, Bennett won and could not be denied the Republican nomination thus destroying fusion efforts. Remaining in the mayoral race as the independent candidate of the wealthy members of the committee of 250, Mitchel steadily lost ground. Election day ended what someone had labeled a "Municipal Revolution Under Mayor Mitchel."[96]

Although Tammany regained New York in 1917 as it had after every other reform spree, the city the Hall regained was not the same one it had lost in 1905 when McClellan split with Murphy. Herein lies the progressive legacy to the good-government cause. Budgeting, accounting, and purchasing systems, the reorganization of the commissioners of accounts office, the Board of Estimate and Apportionment, and borough governments, and the various other changes of like manner were not undone even by Tammany. Progressives, then, had made the first successful attempt to restructure and modernize New York's government. It was left to others to use and build upon this system.

NOTES

1. Quoted in Edward M. Sait, "Research and Reference Bureaus," National Municipal Review, II (January, 1913), 48.

2. Henry Bruere, "New York's Municipal Problems," American Review of Reviews, XLVIII (October, 1913), 471.

3. New York World, April 14, May 16, September 1, 28, 29, October 1, 5, 1905; City Committee Minutes, April 24, May 15, June 12, August 24, September 18, 28, 1905; American Monthly Review of Reviews, XXXII (September, 1905), 273, October, 1905, pp. 399-400, November, 1905, pp. 516-517, December, 1905, p. 644; Irwin Yellowitz, Labor and the Progressive Movement in New York State 1897-1916 (Ithaca, N.Y., 1965), pp. 190-198.

4. Bruere, "Reminiscences," p. 42.

5. Ibid., pp. 19-29; Dahlberg, Bureau of Municipal Research, pp. 12-13.

6. George McAneny, "Reminiscences," CUOHC, p. 7; Edwin R. Lewinson, John Purroy Mitchel; The Boy Mayor of New York (New York, 1965), p. 53.

7. Address by Hon. John Purroy Mitchel, Commissioner of Accounts, at the Budget Exhibit, October 9, 1908, Box 2, Papers of John Purroy Mitchel, Library of Congress; Mitchel to Mayor George B. McClellan, December 31, 1909, Box 3, ibid.; William H. Allen to George McAneny, August 20, 1907, McAneny Papers.

8. Quoted in George B. Hopkins, "The New York Bureau of Municipal Research," Annals, XLI (May, 1912), 239.

9. "Investigation of Office of President of Borough of Manhattan," Box 37, Mitchel Papers; "A Report on a Special Examination of the Accounts and Methods of the Office of the President of the Borough of Manhattan," July 16, 1907, Box 38, ibid.; Hughes Public Papers, 1907, p. 275.

10. Hughes Public Papers, 1907, pp. 275, 280.

11. Dahlberg, Bureau of Municipal Research, p. 19.

12. Hughes Public Papers, 1907, pp. 242-250; "In the Matter of the Charges Against the Hon. Louis F. Haffen, President of the Borough of the Bronx," November 9, 1908, Box 37, Mitchel Papers; typed paper prepared by the Bureau of Municipal Research in re charges against Louis F. Haffen [n.d.], ibid.

13. Quoted in Lewinson, John Purroy Mitchel, p. 44.

14. Address by Mitchel at Budget Exhibit, October 9, 1908, Box 2, Mitchel Papers.

15. Lewinson, John Purroy Mitchel, p. 34.

16. "Results and Limitations of Financial Control," Municipal Research, p. 2; William A. Prendergast, "Efficiency Through Accounting," Annals, XLI (May, 1912), 43-45; "Significant Points Brought Out in the Report of the Bureau of Municipal Research to Comptroller Metz on the Need for Reorganization of the Business Methods of the Department of Finance" [n.d.], McAneny Papers.

17. Cleveland, Municipal Administration and Accounting, pp. 353-354.

18. Henry Bruere, "The Budget as an Administrative Program," Annals, LXII (November, 1915), 177.

19. "Results and Limitations of Financial Control," Municipal Research, p. 1.

20. Bruere, "Reminiscences," pp. 23-25; "The New York City Budget," Municipal Research, No. 88 (August, 1917), pp. 95-96; Dahlberg, Bureau of Municipal Research, pp. 13, 155-158.

21. Quoted in "New York City Budget," Municipal Research, p. 97.

22. Ibid., pp. v, 9, 16, 97-98; Prendergast, "Efficiency Through Accounting," pp. 44-45.

23. Minutes of the Board of Trustees of the Bureau of Municipal Research, February 4, April 8, 1908, McAneny Papers; "Bureau of Municipal Research Report on Need for Reorganization of the Department of Finance" [n.d.], ibid; New

York City Accounts on a Business Basis [n.d.], issued by Bureau of Municipal Research, Papers of Lillian Wald, NYPL; "Results and Limitations of Financial Control," Municipal Research, pp. 3-5.

24. Dahlberg, Bureau of Municipal Research, p. 203.

25. XL (July, 1909), 17.

26. William H. Allen, "New York's First Budget Exhibit," American Review of Reviews, XXXVIII (December, 1908), 686-688.

27. Raymond V. Ingersoll, "An Estimate of Mayor Gaynor," National Municipal Review, III (January, 1914), 78-81; Mortimer Smith, William Jay Gaynor: Mayor of New York (Chicago, 1951), pp. 1-64; Dahlberg, Bureau of Municipal Research, p. 20.

28. American Review of Reviews, XL (July, 1909), 17, (October, 1909), p. 402, (November, 1909), pp. 52-524; The Municipal Campaign [1909], McAneny Papers; Yellowitz, Labor and Progressive Movement, p. 225.

29. Campaign Speech for Office of President of the Board of Aldermen [1909], Box 43, Mitchel Papers.

30. Speech after endorsement of nomination by Committee of One Hundred [1909], ibid.

31. Municipal Campaign, McAneny Papers.

32. American Review of Reviews, XL (November, 1909), 596.

33. Ibid., December 1909, pp. 652-653; Lewinson, John Purroy Mitchel, pp. 51-53.

34. November 3, 1909, Box 3, Mitchel Papers.

35. Henry Bruere, "Efficiency in City Government," Annals, XLI (May, 1912), 5.

36. William B. Shaw, "The Men Who Are Governing New York City," American Review of Reviews, XLI (March, 1910), 301.

37. Prendergast, "Reminiscences," II, 304.

38. Hopkins, "Bureau of Municipal Research," p. 238; Allen, "Reminiscences," I, 113-114.

39. Prendergast, "Reminiscences," II, 313; Allen, "Reminiscences," II, 193-194; Lewinson, John Purroy Mitchel, pp. 65-66.

40. Edmund D. Fisher, "Efficiency in the Fiscal Operations of Cities," Annals, XLI (May, 1912), 71.

41. William A. Prendergast, "New York City Finances," National Municipal Review, II (April, 1913), 223-229.

42. Prendergast, "Efficiency Through Accounting," pp. 44-54; Bruere, "Budget as Administrative Program," pp. 178-179; Fisher, "Efficiency in Fiscal Operations," pp. 76-77.

43. "New York City Budget," Municipal Research, pp. ix-xii; Bruere, "Budget as Administrative Program," p. 180.

44. Bruere, "Budget as Administrative Program," pp. 181-183.

45. Prendergast, "Efficiency Through Accounting," p. 51.

46. Bruere, "Efficiency in City Government," p. 13.

47. Ibid.; Prendergast, "Efficiency Through Accounting," pp. 51-56.

48. Citizens' Municipal Committee of the City of New York, Treasurer's Report for the Campaign of 1913, pp. 3-5, 22-23, 1913 Political Campaign, Charles L. Bernheimer Papers, NYHSL: New York World, March 13, 20, 1913; Prendergast, "Reminiscences," IV, 495.

49. "Tentative Statement of Attitudes of Citizens' Municipal Committee" [1913], 1913 Political Campaign, Bernheimer Papers.

50. Citizens' Municipal Committee, Treasurer's Report for 1913, pp. 12-21, ibid.

51. Lillian Wald to Graham Stokes, June 13, 1913, Minutes of Conference on Mayoralty Candidates, Henry Street Meeting, May 27, 1913, Wald Papers; Louis H. Pink, "Reminiscences," p. 23; "The McAneny General Committee," July 29, 1913, McAneny Papers; Citizens' Municipal Committee, Treasurer's Report for 1913, pp. 5-6, Bernheimer Papers.

52. Prendergast, "Reminiscences," IV, 502-509; Pink,

"Reminiscences," pp. 23-24; Lewinson, John Purroy Mitchel, pp. 82-86.

53. Prendergast, "Reminiscences," IV, 515-518; McAneny, "Reminiscences," p. 6; Citizens' Municipal Committee, Treasurer's Report for 1913, pp. 5-6, Bernheimer Papers.

54. Lewinson, John Purroy Mitchel, p. 87.

55. Quoted in Bruere, "New York's Municipal Problems," p. 465.

56. Mitchel to William J. Taylor, Chairman Independence League, City Committee, August 25, 1913, Box 5, Francis W. Bird to Mitchel, September 25, 1913, Mitchel to James A. Allen, September 26, 1913, Henry L. Stimson to Mitchel, September 26, 1913, Box 6, Mitchel Papers; Lewinson, John Purroy Mitchel, pp. 92-94.

57. Smith, William Jay Gaynor, p. 177; American Review of Reviews, XLVIII (October, 1913), 394-396.

58. Smith, William Jay Gaynor, p. 183.

59. Prendergast, "Reminiscences," IV, 525-526.

60. Speech by Mitchel accepting the Fusion nomination for Mayor of New York City, delivered at Cooper Union on September 29, 1913, Box 45, Mitchel Papers.

61. New York Evening Post, April 25, 1913.

62. Mitchel acceptance speech, September 29, 1913, Mitchel Papers.

63. Statement of the Plans and Purposes of the Citizens' Municipal Committee [n.d.], Bernheimer Papers; Platform for Citizens' Committee, 1913, Campaigns, Box H-4, CU Coll.; Municipal Platform prepared by the Platform Committee of the National Progressive Party in the City of New York for the New York City Campaign of 1913, Wald Papers; Bruere, "New York's Municipal Problems," p. 467.

64. Bruere, "New York's Municipal Problems," pp. 470-471.

65. American Review of Reviews, XLVIII (December, 1913), 647-648; Lewinson, John Purroy Mitchel, p. 100; New York World, November 5, 1913.

66. Henry Bruere, "Mayor Mitchel's Administration of the City of New York," National Municipal Review, V (January, 1916), 37.

67. Bruere, "Efficiency in City Government," p. 5.

68. Bruere, "Mayor Mitchel's Administration," pp. 26-32; Albert Shaw, "New York City's Government by Experts," American Review of Reviews, XLIX (February, 1914), 173-178; Harold Seidman, Investigating Municipal Administration (New York, 1941), pp. 65-66; Bruere, "Reminiscences," pp. 102-104.

69. Bruere, "Mayor Mitchel's Administration," p. 32; Bruere, "Reminiscences," pp. 87, 92-102; Allen, "Reminiscences," II, 188-189; Prendergast, "Reminiscences," III, 458-464; Mitchel to Ogden Mills, April 5, 1915, Box 5, Mitchel Papers; Seidman, Investigating, pp. 66-75.

70. Bruere, "Budget as Administrative Program," pp. 176, 181-185, 187-188; "Results and Limitations of Financial Control," Municipal Research, pp. 8, 11-16; "New York City Budget," Municipal Research, pp. iii-vi, 21-22; "Next Step in the Development of a Budget Procedure for the City of Greater New York," Municipal Research, No. 57 (January, 1915), p. 7.

71. Bruere, "Budget as Administrative Program," pp. 185-186.

72. Ibid., p. 183.

73. Tilden Adamson, "The Preparation of Estimates and the Formulation of the Budget--the New York City Method," Annals, LXII (November, 1915), 249-253; Dahlberg, Bureau of Municipal Research, pp. 207-208.

74. Dahlberg, Bureau of Municipal Research, p. 207.

75. Karl, Executive Reorganization, p. 138.

76. Lewinson, John Purroy Mitchel, p. 128.

77. Bruere, "Mayor Mitchel's Administration," pp. 34-35; John Purroy Mitchel, Presentment of the City's Legislative Program for Home Rule and Financial Relief, January 28, 1916, Box 30, Mitchel Papers; Lewinson, John Purroy Mitchel, pp. 125-127, 129.

78. Sol Cohen, Progressives and Urban School Reform; The Public Education Association of New York City, 1895-1954 (New York, 1964), pp. 87, 94-95; Lewinson, John Purroy Mitchel, pp. 154-160, 162-163, 171.

79. Mitchel to Ogden Mills, April 5, 1915, Box 8, Mitchel Papers.

80. William A. Prendergast, "Why New York City Needs a New School Plan," American Review of Reviews, LII (November, 1915), 588.

81. Ibid., pp. 584-586; Lewinson, John Purroy Mitchel, pp. 163-166; Bruere, "Mayor Mitchel's Administration," p. 33.

82. Lewinson, John Purroy Mitchel, p. 166.

83. Prendergast, "Why a New School Plan," p. 587.

84. Ibid., pp. 586-588; Lewinson, John Purroy Mitchel, pp. 167-169.

85. Dahlberg, Bureau of Municipal Research, p. 240.

86. American Review of Reviews, XLIX (February, 1914), 171-178.

87. Riordan, Plunkitt, p. 25.

88. Garrett, LaGuardia Years, pp. 45, 344, n. 82.

89. Lewinson, John Purroy Mitchel, p. 246.

90. Ibid., pp. 176-188.

91. William H. Allen, "Tuberose Obstacles to Reform in New York City," National Municipal Review, V (July, 1916), 422.

92. Lewinson, John Purroy Mitchel, pp. 137-143, 189-190.

93. Quoted in ibid., p. 200.

94. Ibid., pp. 200-201, 230-236; Raymond V. Ingersoll, "The Recent New York City Fusion Campaign," National Municipal Review, VII (March, 1918), 187-189; Garrett, LaGuardia Years, p. 45.

95. H. S. Gilbertson, "Municipal Revolution Under Mayor Mitchel," American Review of Reviews, LVI (September, 1917), 300; Joseph M. Price, "Fusion Mistakes and a Way Out," National Municipal Review, VII (March, 1918), 183-184; Lewinson, John Purroy Mitchel, p. 217.

96. Lewinson, John Purroy Mitchel, pp. 227-229; Gilbertson, "Municipal Revolution," p. 300.

Chapter 9

THE PUBLIC ADMINISTRATION OF HEALTH, HOUSING, AND UTILITIES

Equipped with new municipal machinery--a health department laboratory, tenement house department, and public service commission--progressives anticipated ushering in a new era in city administration and social services. Accepting the implications of bureaucratic method, they created new divisions, bureaus, and offices, generally along functional lines, that further systematized the operations of the original departments and commissions. Thus New York City's government possessed the means to be involved continuously in health, housing, and utility problems.

Progressive innovations in government concentrated much power in administration. Good government increasingly became dependent upon good administrators. New Yorkers learned, particularly with reference to their public service commission, that even reformers sometimes were poor managers. Tammany Hall, ever interested in maintaining its power in New York, sought to control key administrative posts and when able "Tammanyized" the tenement house department and public utilities commissions. It became clear that good men still were needed if good government were to prevail. Yet few responsible politicians, even those critical of the performance of New York's bureaucrats, suggested dismantling any of the new commissions or departments. By 1920 bureaucratic management of slum housing and utilities had become a permanent feature of New York City's government.

In a mixture of old and new reform, the progressive Hermann Biggs became general medical officer of New York and Lawrence Veiller's tenement house department began its labors under the aegis of Mayor Seth Low in 1902. Low had hoped to appoint Biggs commissioner of health, but the young doctor, his own health in jeopardy and with a family to support, refused to take on this arduous administrative task. As general medical officer, Biggs was free from administrative chores yet involved in the formulation of health department policies. Though serving under Commissioners Ernst J. Lederle and Dr. Thomas Darlington, he was "the real directing force" behind New York's health program. "The New York City Health Department from 1902 to 1913," writes Biggs' biographer, "was in large measure the concrete expression of the mind of Hermann Biggs, and the history of its development is the history of the progress of his sanitary statesmanship."[1]

Biggs and the health department continued to apply laboratory science to public health and formulate programs to control disease. Much of this effort was directed toward the young as the department pioneered in school and child-health services. In 1902 at the urging of Lillian D. Wald it appointed this country's first school nurses and three years later expanded medical examinations of school children to include youngsters with noncontagious diseases. This was "a step of far-reaching importance in the process of enlarging the function of the Health Department from the control of epidemic disease to the promotion of health in its widest aspects." To centralize responsibility for child-health services, a division of child hygiene was created in 1908 after a successful crusade begun by Dr. S. Josephine Baker. Among the new agency's functions were the control and supervision of midwives, the inspection and sanitary supervision of day nurseries, and the medical inspection and examination of school children. Under Dr. Baker's direction the new division increased the number of school nurses, systematized their activities, and established health care and educational centers throughout the city. Further to protect the young, the board of health crusaded for pasteurized milk and established infant milk stations.[2]

Municipal efforts to control and cure tuberculosis accelerated under Biggs' leadership. The department opened a tuberculosis dispensary and hospital and sent nurses to visit homes struck by the disease. In 1905 the city pioneered in the establishment of a sanitarium. The techniques used in the struggle against tuberculosis were then extended to venereal disease: diagnostic laboratory tests, separate hospital facilities, and an elaborate system of reporting.[3]

In 1913 Dr. Biggs resigned from his position as chief medical officer ending an association with New York City that had extended over a period of twenty-six years. "It seems probable," his biographer proudly writes, "that few individuals have ever made more fundamental contributions to the welfare of the city than this same young man who had come to it, knowing only three people, thirty years before."[4] Biggs did not retire from reform but went on to revolutionize New York State's health department and program.

Veiller's tenement house department, incorporated into the city's charter in 1901, was one of New York's earliest experiments in the bureaucratic management of a pressing urban problem. What success the department had in supervising slum housing and fulfilling reformers' expectations was due in large measure to the energy, interest, and vigilance of Veiller, Robert W. deForest, and the tenement house committee of the Charity Organization Society.

Because of his technical knowledge and organizing ability, Veiller was the logical choice to head the new department, but Mayor Low considered him too radical. Instead Low offered the commissionership to deForest who accepted only after Josephine Shaw Lowell pressured him and Veiller consented to serve as first deputy commissioner. As it turned out, Veiller did most of the work. Both men, however, quit the new agency in 1904 when Tammany Mayor George B. McClellan took office. Fortunately for the experiment, Veiller did not abandon housing reform. He kept an eye on the tenement house department and reactivated the tenement house committee of the Charity Organization Society as a private watchdog committee to oversee the new city bureau.[5]

Mayor McClellan replaced deForest with T. C. F. Crain, a Tammany politician, and in just a few months Veiller complained that Crain was destroying the new department. In a memorandum of charges, cleared first with the tenement house committee of the Charity Organization Society and then presented to Mayor McClellan, Veiller accused Crain, "a hypocrite through and through," and the department of inefficiency, corruption, and neglect of duty.[6] Specifically, Veiller charged that Crain neglected "to vacate houses . . . unfit for human habitation," ignored reports from inspectors and thus failed to improve old tenements under the terms of the law of 1901, misused public funds by appointing a horde of unnecessary employees, blackmailed tenement house owners with false charges of violations, and generally neglected to administer the tenement house code.[7] To dramatize his charges, the brilliant and politically astute Veiller took the mayor on a tour of some of Manhattan's worst slums. Impressed, McClellan forced Crain to resign.

Crain's two successors were fairly competent men and restored a measure of prestige to the tenement house department. Edmund J. Butler, long active in various Catholic charities, served as commissioner until the end of McClellan's mayoralty in 1909. Mayor William J. Gaynor appointed John J. Murphy, former secretary of the Citizens' Union, who served through 1917. "Mr. Murphy," Veiller recalled, "was a fine man, entirely honest, not brilliant, with very little initiative but very suave and jovial." Under Murphy's leadership even more than Butler's, the tenement house department gained momentum and seemed honestly and efficiently run. For most of these years Veiller was the unofficial guiding force behind the department.[8]

The problem under Crain and the need for the tenement house committee of the Charity Organization Society to act as a watchdog suggest weaknesses in progressive reform techniques. Good men still were needed for effective government, perhaps even more so in an expanded municipal bureaucracy. And to whom a bureaucracy was responsive was as important as the employment of worthy bureaucrats. The tenement house department, as a new municipal machine, even proved usable to Tammany. As Veiller charged, the Tiger found it a convenient channel for blackmail and a home for needy jobseekers.

Of course, the tenement house rules always could be changed--for better or worse. Most amendments for improvements in the law or enforcement machinery originated either in the tenement house department or the tenement house committee of the Charity Organization Society. Even builders, owners, and others interested in housing who from time to time felt certain provisions of the code were unfair often cleared their proposals with these two agencies. More often, builders, real-estate men, old-law tenement owners, lending institution representatives, and some architects sought to weaken or destroy the code. But the ever vigilant and persuasive Veiller, who remained in Albany for most of each legislative session as agent for the Charity Organization Society's tenement house committee, prevented passage of most of the damaging legislation. Word got around, he later boasted, "that it was hopeless to try and make any changes in the law 'unless Veiller approved it.'" In just the first two years of the tenement law's existence Veiller worked to defeat thirty-one bills.[9]

New York's court of appeals precipitated the tenement house department's severest crisis. In February, 1912, the court found "that an apartment house and a tenement house for purposes of legal regulation were different."[10] The ruling excluded from the department's jurisdiction every house in which each family had its own bathroom, toilet, and kitchen facilities. All of New York's new-law tenements had these facilities and thus under the ruling would be exempt from municipal supervision. To stop even the court from destroying his handiwork, Veiller fought back and in the process created an interesting combination of progressive supporters. He and deForest, supported by the tenement house committee of the Charity Organization Society, conferred with Commissioner Murphy and prepared a bill for the state legislature that amended the tenement house law making it explicit that even apartments with up-to-date facilities fell within the meaning of tenement house in the law of 1901.

Veiller hurried to Albany, cornered Robert F. Wagner, one of Tammany's brighter young men in the legislature, and persuaded him to sponsor the housing amendment. He also convinced Wagner to pressure Democratic Governor John A. Dix into delivering an emergency message to the state's lawmakers and drummed up support among his friends in the Republican-dominated assembly.[11] "My plans were followed and in 24 hours the Bill had passed both Houses and was on the Governor's desk for his signature," related Veiller.[12] After a delay of ten days Governor Dix signed the bill. If this court decision had been permitted to stand, Veiller suggested, "chaos would have resulted in New York City and the results of 50 years' effort at housing reform would in a short time have gone for naught."[13]

Some of the strongest opposition to Veiller's code came from Brooklyn, the borough that labeled itself "a city of homes." Builders who wished to abandon Brooklyn's traditional two-family house, not covered by the tenement house law of 1901, for the more profitable but legally supervised three-family dwelling were joined by financially pinched owners of old but large one-family houses, who hoped to convert their properties into three-unit apartment houses, in pressing the legislature to weaken provisions in the law of 1901 concerning light, air, sanitation, and fire prevention. For fourteen years Veiller managed to ward off these attacks, but in 1916, to his dismay, the Brooklyn Bureau of Charities as well as several members of his own tenement house committee supported the Brooklyn interests. Virtually alone, Veiller denounced his own friends as being "subservient to certain real estate interests," and resigned from the tenement house committee of the Charity Organization Society. Although the determined housing reformer continued to fight, the legislature passed the Lawson bill which permitted in private dwellings converted into three-family houses the dangerous air shaft, inflammable wooden stairs, small, dark, and interior alcoves, and other violations of the law of 1901. From this time, Veiller, the brilliant creator of urban progressivism's answer to housing evils, believing his friends had betrayed him, turned most of his energies outward to the national housing movement.[14]

At the outset in 1901, Veiller organized the new tenement house department and carefully outlined the duties for each bureau. Divided into three major divisions--buildings, inspection, and records--the department was well structured for its task of supervising tenements. The buildings division was responsible for checking the plans for new tenements and regularly inspecting the houses under construction. No building could be occupied until the builder received from the department a license indicating the house met all the requirements of the law. The bulk of the department's routine work fell upon the inspectors who supervised all tenements, old and new. In keeping with the progressive emphasis on securing facts as the basis for remedial action, the records division of the department compiled and classified descriptive data on the tenements and their inhabitants. During the first two years the department had a force of about forty employees. By 1916 the work force had increased to 588 with approximately one-third of the employees engaged in inspection.[15]

For the first two years "a pioneering zeal pervaded the Department's work."[16] Veiller and deForest, who then ran the new agency, optimistically reported after eighteen months that the results of the new housing code had "more than fulfilled the anticipations of the framers of the law."[17] The department had made 337,246 inspections; filed 55,055 violations; investigated 32,825 citizens' complaints; and ordered 21,548 plumbing items repaired. By 1916 thousands more of the old tenements had been inspected and generally made to conform to the standards of the new law. Although over twenty thousand new tenements had been built during this time, the old "dumbbell" had been replaced by a new-law building having more airy rooms and halls, larger interior courts and yards, private water closets in each apartment, and iron fire escapes.

Such success did not mean, as Roy Lubove states, that the department ushered in a "golden age of glistening, white washed tenements." Measured against the enormity and persistence of New York's slum housing problem, the new department's achievements perhaps seem meager indeed. Yet in spite of the magnitude of the task facing the department, more than ever before had been accomplished. On the basis of the department's yearly reports, Lubove records that between 1902 and 1914 the department had answered over 500,000 tenant complaints, filed violations against 193,000 of them, and supervised the construction of over 24,000 new tenements.[18] "On balance," writes Lubove, "the idea of a separate housing department paid off in terms of a cleaner and healthier New York."[19]

New York's government was continuously and directly involved in slum problems; the locus of this concern was not in a legislative body, but in a new permanent executive department. Tenants, landlords, and builders now had to deal with municipal administrators. True, this very early experiment in urban bureaucratic government had many shortcomings. The department was understaffed and underpaid, the turnover in personnel excessive, administrative confusion stalled orders and decisions, tenants and landlords often proved uncooperative, inspections were too irregular, enforcement was uncertain and at times timid, and buildings were frequently occupied before proper licenses had been secured. Nevertheless, its deficiencies involved details and not form. Veiller had created a technique for dealing more effectively with tenements and by implication other urban problems. Progressives fighting other battles looked to the tenement house department as a model for their own efforts at sketching a new relationship between New York City and its citizens.

The crucial question is not how well Veiller's new reform agency functioned, but whether it alone was enough to solve New York's incredible housing situation. The enforcement of restrictive legislation was essentially negative; that is, it could provide minimum standards but could not create more desperately needed housing units. Thus by 1917 even Veiller increasingly was forced to defend what he had created. Newer people, architects, and housing economists argued that restrictive legislation, while necessary, could not alone secure enough low-cost housing for the city's lowest income groups. Moving beyond the father of New York's tenement house code, such reformers demanded government built, owned, and subsidized housing for the poor. This move pointed toward the future and Veiller fought the trend as an unwanted socialism.[20]

On July 1, 1907, New York City's public service commission began its labors. Progressives anticipated that it would "set a new standard for public service," but were apprehensive when few of Governor Hughes' appointees as commissioners were by reputation utility experts. Actually, few men in 1907 were sufficiently experienced to perform the many functions assigned to the commission.[21] However, all of the governor's choices were men "whose integrity was above suspicion and who were not actively connected with the political machines of either party."[22] Like the Bureau of Municipal Research, which, though a private organization, served as a staff agency for New York City's government, the public service commission carried on its supervisory and planning functions through a number of bureaus and divisions which employed engineers, accountants, lawyers, and other experts. Thus equipped, this "elaborate enginery" was to fulfill progressive dreams.[23]

To establish a factual basis for long-range decisions, the public service commission directed its bureau of statistics and accounts and committees on accounting to devise a system of uniform accounts and records for all utilities. The commission's experts conferred with representatives of the Interstate Commerce Commission, the public service commission of the second district in New York, the utility corporations, and private business organizations and in 1909 began installing prescribed accounts in all rapid transit, street-railway, and gas and electric corporations. These records, similar in design and purpose to those installed in city departments by Comptrollers Metz and Prendergast, were meant to secure financial honesty by showing clearly and accurately the source of all income and purpose for each expenditure. They became the basis for the public service commission's publication of the financial operating statistics of each corporation and decisions regarding corporate requests for permission to issue new stock.[24] While of unquestioned value, this task early in the commission's career combined with the franchise bureau's classification of all extant utility franchises consumed much time, buried the commission in statistics, and projected the agency not as a dynamic regulating body but as something of an information center. Just as the economy and efficiency efforts of the Bureau of Municipal Research, the fusion Board of Estimate, and Mayor Mitchel were not substitutes for the granting of social services needed by New Yorkers, better accounting systems for utility corporations hardly relieved transit congestion, lowered rates, or increased the quality of gas and electricity. And these were matters uppermost in the minds of New Yorkers.

Still, by 1911 when Hughes traded the governor's seat for a position on the United States Supreme Court, the public service commission had settled into a routine of work and had achieved modest success in attacking some of the more obvious and longstanding utility abuses. Over a period of several years the commission ordered a number of improvements in transportation service, equipment, and safety devices. The surface lines in Manhattan and the Bronx were ordered to overhaul and repair all cars making them cleaner, safer, more efficient, less noisy, and generally more comfortable. Wheel-guard fenders, guard rails, and other safety devices became standard equipment at the commission's prodding. In fact, the commissioners required that every accident connected with New York's transportation facilities be reported to its office. The rapid transit and elevated lines were forced to install new signal and brake systems, side doors on cars, and better lighting facilities. New stations were built and many station platforms were lengthened to accommodate longer trains. To increase service, especially during the rush hours, the commissioners ordered that more cars be operated on all the lines, the frequency of arrival increased, and rush-hour service begun earlier and continued later.

Because gas and electricity rates were fixed in the law that created the public service commission, the commissioners felt no compulsion to investigate the reasonableness of the eighty-cent gas rate and ten- to twelve-cent electricity rate. They also failed to take the initiative on other service matters connected with the almost thirty gas and electric companies under their jurisdiction and waited instead for complaints from consumers. This was most surprising in light of the decades-long public agitation against the gas and lighting trust. Among the typical complaints investigated and informally handled by the commission were those relating to poor quality or inadequate supply of gas or electricity, unfair contract stipulations, rate discrimination, erroneous billing, leaky pipes and connections, the failure to provide service, arbitrary discontinuance of supply, and the removal of meters without notification. In settling these and other matters the commission encouraged simpler and more standardized contract forms and services. By acting primarily upon complaints, and thus in a disconnected fashion, the commission forfeited the chance to establish guidelines and make clear a minimum of acceptable practice touching all aspects of utility services and conduct. Like the efficiency-minded progressives in New York's government, the utility commissioners seemed more concerned with administrative form and technique than with the citizens' welfare.

The commission's most notable achievement in the gas and electricity field was its systematic inspection and testing of all new and repaired gas and electricity meters. Before the public service commission law only a few meters bearing the state inspector's stamp had actually been tested. Besides, the affixed stamp in no way interfered with the opening and repairing of meters and thus gave official cover to thousands of defective measuring devices. The new commission placed the seal in a way that required its breaking when the meter was opened thus necessitating a new inspection. During its first six months the public service commission tested almost 120,000 gas meters and by 1910 had tested over one million.[25]

The commission had hit at the outer fringes of corporate policy and practice, but had not yet subordinated private utility power to public management. So bogged down in the routine details of its work, the commission rarely showed the imagination or the will to take the initiative in formulating public utility policy. Generally reluctant to wield a big stick, the commission preferred settling most utility service matters informally--by letter, telephone call, or personal consultation. While congenial to corporate executives and well suited to bureaucratic management, such easy, flexible procedures made little public impact.

In its report for 1908 the public service commission boasted that "the principal advantages to be derived from the establishment of a supervisory authority is that it provides a forum to which the consumer and the company may come and secure an adjustment of the difficulty from an impartial and unbiased authority." Such bureaucratic blandness may have prompted the utility corporations "to encourage a complainant to put his case before the Commission," but hardly was calculated to convince the utility corporations that "the operation of public utilities is public business." Moreover, utility consumers expected their commission to function as an advocate, not an impartial judge, as a governing agency, not an information bureau. Predictably, then, an overexpectant public scarcely appreciated the public service commission's real though modest accomplishments or understood the commission's inexperience with public regulation of a giant industry.[26]

Perhaps public impatience stemmed more from a change in perception among Gothamites than from the absence of concrete achievements. Their demands and expectations now were informed by the obvious success of New York's first subway and elevated lines and the equally obvious gap between services and needs. Now it was not a question of some gas or electricity service where none had existed but of insufficient and inadequate service. Also, by the beginning of the century's second decade progressives who crusaded for health, housing, and public administrative reform were consolidating victories, integrating and expanding their programs, and generally pressing for more. By contrast, the initial achievements of the public service commission seemed meager indeed. The commissioners' work seemed out of step with the accelerated pace of reform. Increasingly after 1910 public as well as official criticism of the public service commission grew more intense. Much of it stemmed from the city's desperate need for more transportation.

New Yorkers expected almost instant transit relief when the commission began its labors. And from the outset the commission devoted approximately 60 percent of its budget to transportation construction matters. It supervised completion of the two original subway contracts and approved, after some modification, the former rapid transit board's plans for a Fourth Avenue subway in Brooklyn and a loop line connecting the Manhattan ends of New York's bridges spanning the East River. The commissioners considered other possible subway routes and encouraged the private Manhattan and Brooklyn elevated roads to extend their lines, but to the dismay of Gothamites began little new construction. In a mild slap at the commission, both the Tammany and fusion forces during the mayoral election of 1909 promised more subways.[27]

The public service commission was not deaf to public appeals for better service or blind to the almost inhuman congestion on existing lines. "Present transit conditions are indecent, and the present facilities are wholly inadequate, and yet transportation needs are growing, so that the situation will grow worse until new lines are constructed," the commissioners wrote in their report for 1908.[28] But the city's constitutional debt limit and the rapid transit act of 1891 and subsequent amendments severely restricted the options available to the commission for negotiating new subways or elevated lines. The referendum of 1894 seemed to preclude construction of new subways with private funds, and in 1908 city officials warned the commission that the municipal debt was perilously close to the constitutional limit of 10 percent of the assessed valuation of New York's real estate. Besides, since 1904 the Board of Estimate and Apportionment had to approve all municipal appropriations and bond issues for transit construction. Private funds could be used to extend the existing elevateds, but the commission possessed no leverage with which to persuade the Interborough Rapid Transit Company interests to cooperate. The Elsberg rapid transit amendments, passed in 1906 at the insistence of New York's reform organizations, further limited the commission's negotiating flexibility by fixing in law the duration of contracts for the construction and operation of subways and the financial obligations of the contractor.[29]

Predating the creation of the public service commission, these restrictions were inconsistent with the state's venture into bureaucratic management of its utilities. "With a permanent tribunal to compel adequate service at reasonable rates," wrote Milo R. Maltbie, one of the abler commissioners, "there is no need for the elaborate terms and conditions with which many franchise grants have been loaded down. . . ."[30] Thus, to free itself to deal with any contingency and to achieve greater administrative autonomy, the public service commission worked to modify the rules governing transportation construction. By 1910 it had secured several legislative triumphs, but for an impatient, outraged, and unsympathetic public few new facilities.

In 1908 the commission cooperated with Mayor McClellan's committee on finance and taxation, City Comptroller Herman A. Metz, representatives of the City Club, Citizens' Union, Merchants' Association, and the Board of Trade and Transportation to push through the legislature a constitutional amendment exempting self-sustaining rapid transit bonds from the 10 percent debt limit. The Chamber of Commerce, whose members by this time were hopelessly divided on the issue of public financing of business enterprises, refused to lend its support. In 1909 the legislature again passed the amendment, and in the election that fall it was ratified by the people. The following year, at the urging of the Citizens' Union and Mayor Gaynor's corporation counsel, the state's lawmakers passed legislation enabling the city to benefit from the additional borrowing capacity permitted under the new constitutional amendment.[31]

In the meantime, the public service commission, uncommitted to municipal financing of rapid transit and aware that a minimum of three years was needed to get a constitutional debt-limit amendment, earnestly advocated that private capital "upon payment of part of the earnings to the city or other proper terms" be allowed to construct and operate rapid transit lines. "There is no good reason," the commissioners wrote, "why the city of New York should be cut off from the exercise of both methods of building rapid transit roads." The commission also urged that the city reserve to itself "the privilege to purchase at any time after a certain period, not more than twenty or twenty-five years, and without any payment for the franchise itself." The indeterminate franchise principle, the special cause of Commissioner Maltbie, attached to the cause of private enterprise was successfully incorporated into law in 1909. Under the rapid transit amendment the city could terminate a franchise granted to a private corporation at any time after ten years. In such an event, the city was obligated to buy the property for an amount not exceeding the cost plus 15 percent.[32]

Properly used, this indeterminate franchise was just the

kind of flexible, discretionary, and open-ended managerial power latent in the commission form of regulation. "This principle, for which the Public Service Commission stands, means that the city shall always and continuously be in a position to control transit development," explained Maltbie. In contrast, under a fixed contract the city lost control for a definite number of years. Ever mindful of the rights of the utility corporations, Maltbie further suggested that such a franchise "would not only give better control and protection to the city, but also be fairer to the operator." With the passage of the indeterminate franchise amendment, Maltbie, speaking for the commission, explained that the rapid transit law then represented "a practically harmonious policy which is probably more modern and progressive than any heretofore formulated into law."[33]

The rapid transit legislative package of 1909 also included an amendment extending the indeterminate principle to contracts let for operation of lines only. To increase further the commission's options, another amendment allowed it to finance new construction by assessing the property directly benefitted by new transportation. This proposal was promoted chiefly by the City Club, and the bill was worked out in conferences between representatives of the Chamber of Commerce, City Club, and Maltbie of the public service commission.[34]

By 1909 demands for new transit facilities dominated public discussion and with the rapid transit amendments the public service commission had few excuses for inaction. Indicative of the progressive triumph was that this administrative agency now was the key force in rapid transit development and a center of political controversy. The main issue facing New York and the commission was whether to grant to the existing Interborough lines additional privileges or build a separate municipally owned subway system. In either case, the traction interests, politicians, and other interested parties now had to deal with hired commissioners and not the state's legislators. Thus when the new Gaynor-Fusion administration moved into City Hall in January, 1910, it immediately had to act on an Interborough Rapid Transit Company proposal to build new lines then pending before the public service commission.

To participate effectively in the decision-making, the Board of Estimate and Apportionment appointed a transit committee consisting of Mayor Gaynor, Comptroller William Prendergast, and President of the Board of Aldermen, John Purroy Mitchel. Together with the public service commission it held a series of conferences with representatives of the Interborough Company. These efforts failed, and the transit committee and public service commission subsequently worked out plans for additional subways linking the Bronx, Manhattan, and Brooklyn--the triborough system--and advertised for bids to build the entire line with private funds or certain sections with city money. No satisfactory bids were received.

Only after William McAdoo of the Hudson and Manhattan Railway Company offered to equip and operate certain portions of the triborough system if built with city money did the Interborough Rapid Transit Company again submit a proposal. It offered to extend its existing subway to a size nearly equal to that in the triborough plan, but its price was a fifty-year lease and a construction contribution by New York City of fifty-three million dollars. The public service commission recommended a slightly modified form of this proposal to the Board of Estimate and Apportionment. Mayor Gaynor was pleased, but Prendergast and Mitchel, even then promoting economy and efficiency in city government, vehemently opposed the scheme and complained that its acceptance meant a betrayal of the fusion platform of 1909 that had promised future subways owned, built, and if necessary operated by the city. In particular, Mitchel disliked the fifty-year lease provision and other financial arrangements that unnecessarily promised excessive profits to the Interborough.

To gain support the mayor exploited widespread business antipathy to public construction of rapid transit and fear that it would lead to municipal operation. He encouraged the leaders of the Chamber of Commerce and Merchants' Association to appoint a citizens' committee to study the transit issue. Headed by former Mayor Seth Low and composed of influential men in engineering, finance, real estate, and commerce, this citizens' committee predictably urged the city to accept the Interborough Rapid Transit Company's offer. Unpersuaded, the fusion Board of Estimate and Apportionment rejected the proposal but also failed to support municipal construction.

To renew the quest for more transportation the Board of Estimate and Apportionment appointed a new transit committee composed of Borough Presidents George McAneny, Cyrus Miller, and George Cromwell. McAneny quickly dominated the group. In March, 1911, the McAneny committee received new proposals from the Brooklyn Rapid Transit Company and the Interborough Rapid Transit Company. Not satisfied with these proposals the borough presidents conferred further with both transit companies and issued a joint report outlining a comprehensive plan for rapid transit development. Labeled the dual plan, it laid out routes for elevated and subway lines for both the Interborough and Brooklyn rapid transit companies and provided for a sharing of construction costs between the city and the operating companies. The Board of Estimate unanimously accepted this transit scheme, although Mitchel questioned some of its provisions.

The Brooklyn Rapid Transit Company accepted its lines, but the Interborough Rapid Transit Company unexpectantly rejected the terms and instead made a counteroffer that increased its margin of financial protection. Over the protests of Commissioners Maltbie and Mitchel, the public service commission and McAneny committee consented to the Interborough Rapid Transit Company's new demands. However, the Board of Estimate rejected the terms and instead offered the entire contract to the Brooklyn Rapid Transit Company. At this juncture in the subway negotiations, spokesmen for the Pennsylvania Railroad Company and J. P. Morgan and Company joined Seth Low to reopen negotiations on behalf of the Interborough

Rapid Transit Company with the public service commission and Board of Estimate.

Talks dragged on until February, 1912, when the Interborough finally submitted a new offer to the city. Although it was only slightly more favorable to New York than that company's previous offers, both the public service commission and McAneny committee accepted it. Maltbie and Mitchel again dissented. These progressives differed from their fellow reformers on the Board of Estimate and utilities commission not on ideological grounds but in business judgment. President of the Board of Aldermen Mitchel complained that the contracts were "a bad business bargain for the city." These two pressed for further changes in the contracts which delayed final approval by the commission and full Board of Estimate until March, 1913. Even then Mitchel, Maltbie, and Commissioner J. Sergeant Cram refused to vote for the final contracts.[35]

Throughout the negotiations Mitchel consistently argued for a municipally owned subway system. In this he was supported vigorously by the Hearst press and, as Mitchel's biographer points out, earned Hearst's gratitude and invaluable support in the mayoral election of 1913. But unlike Hearst, who supported the public ownership of all utilities and other measures designed to attract lower middle- and working-class votes, Mitchel examined the subway issue as a businessman would a new investment opportunity. In all of his political campaigning Mitchel had promised the voters a business government that would "secure a dollar's worth of result for every dollar spent." The subway contract, he believed, did not do that. When Mitchel could not get a municipally owned subway as the most efficient way to secure additional rapid transit, he consistently carried on his fight to get the best terms from a private company for the city. Thus his opposition to the dual subway contracts centered on the excessive profits guaranteed to the companies, the slim chance that the city would ever receive a share of the revenue, the city's obligation to assume financial responsibility should the new road fail to make a profit, and other financial and technical matters. He differed from Prendergast and McAneny, who supported the dual contracts, only in matters of business judgment. These efficiency-minded progressives were more interested in the means by which New York secured transportation than in the service itself.[36]

Mitchel and Maltbie faced stiff opposition from the business community. Led by the Merchants' Association, delegates from over fifty commercial and civic organizations journeyed to Albany in support of the dual plan. Tax conscious businessmen argued that New York did not have the money to build and operate its own system. They also rallied behind the dual subway contracts as a bulwark against municipal operation and socialism. Businessmen had little to fear, however. In the main the public service commission and McAneny committee were hostile to the notion of a wholly city-owned, built, and operated subway system and even refused to threaten such a system to gain leverage in bargaining with the private utilities.[37] Because of this timidity, the number of years the negotiations dragged on, the controversy surrounding the contracts, and the widespread belief that the contracts constituted "the most stupendous local franchise bargain that has ever been considered by any city in the world," the commission gained little public esteem for its labors. In a critique of the contracts in the National Municipal Review, Delos F. Wilcox, a severe critic of the public service commission, suggested that New York's official negotiators, "so solicitous for the protection of private profits," perhaps were "still thinking in the grooves or 1875."[38]

In 1911, while the transit negotiations were in progress, the Democrats gained control of the governor's office and Tammany, generally on good terms with the utility corporations, again exercised power at the capital. When the extremely capable Edward Bassett's term expired in June, 1911, Democratic Governor John A. Dix replaced him with Tammanyite Cram, whose chief distinction in New York, Delos F. Wilcox bitterly wrote, "was that he had served for many years as the social mentor of Boss Murphy."[39] When McCarroll's term ended Dix appointed George V. S. Williams, a Brooklyn lawyer, regular Democrat, and former member of the legal staff of a Brooklyn rapid transit company. Though a man of ability, Williams viewed utility problems too much like the utility corporations to please some progressives. Governor William Sulzer, in spite of urgent pleas to the contrary, replaced Chairman William R. Wilcox at the end of his term in 1912 with Edward E. McCall, a loyal son of Tammany and since 1902 a judge of the state's supreme court. Thus in six years the public service commission not only had a Tammany chairman, but a Tammany majority. The Tiger was in charge of the government's utility machine just when millions of dollars of rapid transit contracts were being awarded by the city. The next year Governor Martin H. Glynn filled a vacancy with Robert Colgate Wood, a Bronx "gentleman who had some years previously served as president of a street railway company." "What the governors of the state of New York since Hughes have done to the public service commissions," lamented Delos Wilcox, "is enough to make the angels weep."[40]

By 1914 what the progressives had built was in alien hands and the object of bitter public distrust. "Generally speaking, however, the popular prestige of public service commissions is waning," wrote Delos Wilcox that year.[41] "The alleged Tammanyizing of the New York state commission under Governors Dix and Sulzer" has destroyed its prestige, wrote another critic. New York's experiment had become an example "of the possibilities for evil that the system contains."[42] "The Crisis in Public Service Regulation in New York" was the title of an article by Delos Wilcox a year later.[43]

In response to the crisis a joint committee of the state legislature investigated not the utilities as in the past but New York's two public service commissions. Its report was followed by the appointment of a second committee that in 1917 recommended the "enginery" be overhauled. "As an agency of the State for real regulation and supervision of public utilities," the investigating committee concluded, "the Public Service Commissions of this State have absolutely failed from their organization to the time of the beginning of this investigation."[44] This indictment of New York's regulatory experiment was based on an exhaustive and broad inquiry into the utilities commissions' organization, proceedings, and transactions. The committee examined the commissions' exercise of power and administration of the utilities law and the conduct and activity of the commissioners and staff. Special attention was given to New York's rapid transit facilities, rates, and services and the dual subway contracts.

The legislators found that the Tammany commissioners generally were incompetent, unfamiliar and unconcerned with their tasks, dilatory in the performance of their duties, and reluctant to inconvenience the utility corporations. The commissioners spent too much time in private business pursuits, had previous and even current connections with utility corporations, used official cars for personal travel, and engaged in other unethical activities. During the course of the investigation four of the commissioners for the first district were swept out of office through resignation, removal, or indictment.[45] Little wonder the state's lawmakers concluded that "the whole scheme of regulation had failed in New York." They recognized that "an experiment in method," the administrative device, had ultimately become "an experiment in men."[46]

The state's legislators severely criticized the public service commission's excessive reliance upon consumer complaints before acting against utility abuses, inefficient methods of handling complaints, unwillingness to enforce its own orders, lack of decisiveness in policy-making, failure to assume responsibility, unscientific approach to problems, and general procrastination in all of its supervisory activities. The investigators accused the commission of disregarding the public interest and making excessive and unnecessary commitments to the railroad companies in the dual subway contracts. "The contract finally accepted," the committee concluded, "was the most disadvantageous to the city of all of the many that were offered during the years preceding negotiations for the operation of these extended franchises." Even the subsequent letting of the contracts "was careless and improvidential, and through them the city lost several million dollars," the lawmakers argued.[47]

Not content just to describe wrongs, the state's investigating committee placed much of the blame for the commission's failure as a regulatory body on its activity as a construction agency. The committee strongly urged that New York's traction problem be separated from other aspects of utility regulation. Essentially, the investigators were asking for the re-creation of a rapid transit board for New York City. The legislature responded favorably to this suggestion, and by 1921 New York had a single state-wide public service commission with jurisdiction over gas, electricity, steam, telephone, telegraph, and transportation corporations. A separate transit commission was created with broad powers to regulate and construct New York City's transportation facilities. By this date, too, the personnel of the commissions had undergone several turnovers.[48]

Thus New York's first experiment in bureaucratic management of all of its public utilities ended under a barrage of criticism. The commissions' apparent failures stemmed from the nature of urban progressivism, the city's desperate need for more transportation, and the realities of New York politics. The progressives placed too much faith in techniques, methods, and machinery and too often forgot that the men needed to run the machinery were also important. Furthermore, the reformers forgot that creating a method to deal with a problem was not identical with solving the problem. In other words, good bureaucrats were needed to make a good bureaucracy. And to whom the bureaucracy was responsive was also vital. That lesson in administrative reform needed to be learned. Progressives also forgot that even experts, reformers, and other potential public servants shared foibles common to all men. The progressives' dream of disinterested, objective men using scientific means to arrive at just solutions was just that--a dream. Even experts and seemingly impartial public servants differed on various issues and were influenced by community pressures.

But to emphasize the shortcomings in this experiment is to miss the essence of the progressive triumph. Even the state legislative committee investigating the public service commission understood that "while the experiment of regulation and supervision tried has not justified expectations the policy of regulation and supervision has not had a fair trial, and it is hoped that the scheme of revision and reorganization submitted comprehends such changes in the law as are desirable."[49] There was no suggestion of going back to a pre-1907 arrangement in dealing with utilities. The progressive espousal and creation of bureaucratic means to manage a large portion of New York City's economy was not rejected in the reorganization of the public service commissions but was bequeathed to the next generation. Although new commissions were organized and new men ran them, the progressive method, so well described by Charles Evans Hughes back in 1907, remained intact. "But to prophesy the final outcome of the experiment would now be hazardous," wrote Henry Bruere in 1908.[50] Little did he realize then how hazardous it indeed would be.

NOTES

1. Winslow, Contribution of Hermann Biggs, p. 17.

2. Winslow, Life of Hermann Biggs, pp. 185-187, 193, 215, 222, 226, 229.

3. Ibid., pp. 185-188, 194, 200, 226, 235.

4. Ibid., p. 250.

5. Veiller, "Reminiscences," I, 38-40, 45, 57-58.

6. Ibid., pp. 57-60.

7. "Memorandum in Regard to Tenement House Department Administration," December 24, 1904, Veiller Papers.

8. Veiller, "Reminiscences," I, 64-70.

9. Ibid., pp. 45, 50-51; Lubove, Progressives and Slums, pp. 166-167.

10. Veiller, "Reminiscences," I, 76.

11. Ibid., pp. 74-80; Lubove, Progressives and Slums, pp. 169-170; Report of the Committee on Legislation of the Citizens' Union, 1912, pp. 10-11; Searchlight, II (March 26, 1912), 27; Thirtieth Annual Report of the Charity Organization Society of New York City, 1912, pp. 15-16.

12. Veiller, "Reminiscences," I, 79.

13. Ibid., pp. 76, 80.

14. John Ihlder, "Old Dwellings for New," National Municipal Review, VI (November, 1917), 688-692; Lubove, Progressives and Slums, pp. 171-174.

15. Lubove, Progressives and Slums, pp. 156, 159.

16. Ibid., p. 159.

17. DeForest, "Introduction," I, xiv.

18. Ibid., I, xiv-vi; Robert W. deForest, "Recent Progress in Tenement-House Reform," Annals, XXIII (March, 1904), 108, 111; Lubove, Progressives and Slums, pp. 158-166.

19. Lubove, Progressives and Slums, p. 159.

20. *Ibid.*, pp. 158-166, 177-184.

21. Bruere, "Public Utilities Regulation," p. 17; Delos F. Wilcox, "The Crisis in Public Service Regulation in New York," *National Municipal Review*, IV (October, 1915), 550-551; Walker, *Fifty Years*, p. 209.

22. Wilcox, "Crisis in Regulation," p. 550.

23. Bruere, "Public Utilities Regulation," p. 3; Robert H. Whitten, "The Public Service Commissions of New York," in *The Regulation of Municipal Utilities*, pp. 343-371.

24. State of New York, *Annual Report of the Public Service Commission, First District*, 1908 (Albany, 1909), I, 17, 157-166; 1909 (Albany, 1910), I, 112-116 (hereafter cited *PSC Report*); Milo R. Maltbie, "The Fruits of Public Regulation in New York," *Annals*, XXXVII (January, 1911), 188; Bruere, "Public Utilities Regulation," pp. 14-17.

25. Maltbie, "Fruits of Public Regulation," pp. 173-184; *PSC Report*, 1907 (Albany, 1908), I, 20-26, 127-132; 1908, I, 9-10, 14-15, 18-22, 135-140, 144-147; 1909, I, 8-10; 1910 (Albany, 1911), I, 7-8, 21; 1911 (Albany, 1912), I, 9-12.

26. *PSC Report*, 1908, I, 14, 143; Maltbie, "Fruits of Public Regulation," p. 178; Lyman Beecher Stowe, "How New York Deals with Her Public Service Companies," *American Review of Reviews*, XLII (August, 1910), 211-212; Bruere, "Public Utilities Regulation," p. 10; John S. Kennedy, "The New York Public Service Commissions," *Forum*, XLVIII (November, 1912), 585-586; Fitzpatrick, *Experts in City Government*, p. 31; Wilcox, "Crisis in Regulation," p. 553.

27. *PSC Report*, 1907, I, 9-12; 1908, I, 8, 11-14, 25; 1909, I, 10-12; Maltbie, "Fruits of Public Regulation," p. 189; Lewinson, *John Purroy Mitchel*, p. 67.

28. *PSC Report*, 1908, I, 36-37.

29. *Ibid.*, I, 45-51.

30. *Ibid.*, I, 186.

31. *Ibid.*, 1907, I, 52-53; 1908, I, 39; 1911, I, 8; Maltbie, "Fruits of Public Regulation," p. 190; *CC Annual Report*, 1908-1909, pp. 113-142; *BTT Minutes*, January 8, February 19, 1908;

February 10, March 10, 1909; Minutes of the Board of Trustees of the City Club, February 1, 1906; April 24, 1907; April 15, 1908; Minutes of the Legislative Committee of the City Club, April 11, 1907; Merchants' Association, Bulletin, February, 1908, p. 7; Citizens' Union, Report of the Committee on Legislation, 1908, pp. 9-10; 1909, pp. 15-16; 1910, pp. 16-20; Minutes of the Citizens' Union Committee on Legislation, March 25, April 15, 29, 1910; J. O. Hammitt to Cyrus E. Sulzberger, February 15, 1910, Box 4, Mitchel Papers.

32. PSC Report, 1907, I, 52-53; 1908, I, 39; 1909, I, 18; Milo R. Maltbie, "A Rapid Transit Policy for Greater New York," in The Regulation of Municipal Utilities, pp. 125-127.

33. Maltbie, "Rapid Transit Policy," pp. 124-127.

34. PSC Report, 1908, I, 29-40; 1909, I, 18-23; Minutes of the Board of Trustees of the City Club, February 24, April 12, 21, May 26, 1909; Minutes of the City Club Conference on Proposal to Build Rapid Transit Lines by Assessment on Property Benefited, February 10, 1909; Annual Report of the City Club, 1909, pp. 7-8.

35. Lewinson, John Purroy Mitchel, pp. 67-71; PSC Report, 1909, I, 23-40; 1910, I, 10-13, 47-50; "Complete Report of the Joint Legislative Committee Appointed To Investigate the Public Service Commissions," New York Senate Documents, No. 41, XXI (140 Sess.), 1917, pp. 24-29 (hereafter cited "Committee To Investigate the Public Service Commissions"); National Municipal Review, I (April, 1912), 261-262; McAneny, "Reminiscences," pp. 2-3; Prendergast, "Reminiscences," III, 2, 336-337, 471-472; "President Mitchel's Record in the Matter of Rapid Transit" [n.d.], Box 43, Mitchel Papers; "Report by Seth Low, Chairman, to the Citizens' Committee on Rapid Transit," June 20, 1911, "Report to the Board of Estimate and Apportionment by Special Committee on Pending Transit Proposals," May 21, 1912, Box 116, Low Papers.

36. John Purroy Mitchel to Public Service Commission of the First District, January 20, 1913, Box 5, Mitchel Papers; Delos F. Wilcox, "The New York Subway Contracts," National

Municipal Review, II (July, 1913), 385-386, 390-391; Lewinson, John Purroy Mitchel, p. 71; "Committee To Investigate the Public Service Commission," pp. 50-90; Mitchel acceptance speech, September 29, 1913, Mitchel Papers; Yellowitz, Labor and Progressive Movement, pp. 202-203.

37. Merchants' Association, Bulletin, February 3, 1913, pp. 1-3; February 10, 1913, pp. 1-2; February 17, 1913, p. 1; CC Annual Report, 1913, p. 248.

38. Wilcox, "Subway Contracts," pp. 377, 391.

39. Wilcox, "Crisis in Regulation," p. 552.

40. Ibid., pp. 547, 554-558.

41. Delos F. Wilcox, "Effects of State Regulation upon the Municipal Ownership Movement," Annals, LIII (May, 1914), 75.

42. Stiles P. Jones, "State Versus Local Regulation," Annals, LIII (May, 1914), 94.

43. National Municipal Review, IV (October, 1915), 547-563.

44. "Committee To Investigate the Public Service Commissions," p. 138.

45. Ibid., pp. 16-18, 285-313.

46. Ibid., pp. 9, 21.

47. Ibid., pp. 9-10, 24, 50-84, 103-117, 120.

48. Ibid., pp. 122-130, 135-138, 149; Mosher, "Public Utilities," pp. 247-250.

49. "Committee To Investigate the Public Service Commissions," p. 12.

50. Bruere, "Public Utilities Regulation," p. 17.

BIBLIOGRAPHY

The important primary sources for this study fall into three categories: manuscript collections and public documents, publications of organizations, and contemporary journals, books, newspapers and periodicals. Of the many manuscript collections examined, the most useful were the Papers of Charles L. Bernheimer, Charles E. Hughes, Seth Low, George McAneny, John Purroy Mitchel, Lawrence Veiller, and Richard W. G. Welling; the Reminiscences of William H. Allen, Henry Bruere, George McAneny, William A. Prendergast, and Lawrence Veiller; and the Collections of papers and minutes of the Citizens' Union, City Club, and New York Board of Trade and Transportation. The Documents of the Assembly and Senate of the State of New York were an important source for copies of laws and reports of commissions and boards created during the years under study. Several reports contained useful historical summaries.

The printed annual and special reports, yearbooks, and pamphlets of the Bureau of Municipal Research, Chamber of Commerce, Citizens' Union, City Club, and Merchants' Association were indispensable for an analysis of the people behind many of the reform causes. These sources were supplemented by extensive coverage of city and state affairs in the New York World. On a variety of issues ranging from politics to highly complicated accounting reforms, the American Review of Reviews, the Annals of the American Academy of Political and Social Science, Harper's Weekly, and Municipal Research provided a wealth of material. Books and articles written by participants were valuable not only for descriptions of events but also for an understanding of the reformers' motives, ideals, and goals.

Manuscripts

Reminiscences of George W. Alger. Columbia University Oral History Collection.

Reminiscences of William H. Allen. Columbia University Oral History Collection.

Reminiscences of Joseph C. Baldwin. Columbia University Oral History Collection.

Reminiscences of William S. Bennet. Columbia University Oral History Collection.

Papers of Charles L. Bernheimer. New York Historical Society.

Reminiscences of Robert S. Binkerd. Columbia University Oral History Collection.

Papers of Richard R. Bowker. New York Public Library.

Reminiscences of Henry Bruere. Columbia University Oral History Collection.

Reminiscences of Charles C. Burlingham. Columbia University Oral History Collection.

Papers of Silas A. Burt. New York Historical Society.

Papers of William A. Chanler. New York Historical Society.

Reminiscences of Richard S. Childs. Columbia University Oral History Collection.

Citizens' Union. Campaign [1897] and Shepard-Low Campaign, 1901. Material collected by Everett P. Wheeler. New York Public Library.

Citizens' Union Collection. Columbia University Library.

City Club of New York Collection. New York Historical Society.

Papers of the City Club of New York. New York Public Library.

Minutes of the City Reform Club of New York. New York Public Library.

Papers of William Bourke Cockran. New York Public Library.

Reminiscences of Frederic R. Coudert. Columbia University Oral History Collection.

Reminiscences of Bertram D. Cruger. Columbia University Oral History Collection.
Reminiscences of Samuel Dickstein. Columbia University Oral History Collection.
Papers of Charles S. Fairchild. New York Historical Society.
Reminiscences of Homer Folks. Columbia University Oral History Collection.
Papers of Thomas F. Grady. New York Historical Society.
Reminiscences of John T. Hettrick. Columbia University Oral History Collection.
Papers of Charles E. Hughes. Library of Congress.
Reminiscences of Samuel S. Koenig. Columbia University Oral History Collection.
Mayoralty Papers of Seth Low. Municipal Archives, New York City.
Papers of Seth Low. Columbia University Library.
Papers of George McAneny. Princeton University Library.
Reminiscences of George McAneny. Columbia University Oral History Collection.
Papers of George B. McClellan, Jr. Library of Congress.
Papers of John Purroy Mitchel. Library of Congress.
Papers of Levi P. Morton. New York Public Library.
Papers of the New York Board of Trade and Transportation. New York Historical Society.
People's Institute Collection. New York Public Library.
Reminiscences of Louis H. Pink. Columbia University Oral History Collection.
Reminiscences of William A. Prendergast. Columbia University Oral History Collection.
Reminiscences of Lawson Purdy. Columbia University Oral History Collection.
Papers of Lemuel E. Quigg. New York Public Library.
Minutes of the Reform Club of New York. New York Public Library.
Papers of Jacob A. Riis. Library of Congress.

Reminiscences of Beverly R. Robinson. Columbia University Oral History Collection.

Papers of Theodore Roosevelt. Library of Congress.

Reminiscences of William J. Schieffelin. Columbia University Oral History Collection.

Papers of Gustav Scholer. New York Public Library.

Reminiscences of Laurence A. Tanzer. Columbia University Oral History Collection.

Papers of Lawrence Veiller. Columbia University Library.

Reminiscences of Lawrence Veiller. Columbia University Oral History Collection.

Reminiscences of William H. Wadhams. Columbia University Oral History Collection.

Reminiscences of James W. Wadsworth. Columbia University Oral History Collection.

Papers of Lillian Wald. New York Public Library.

Diary of Richard W. G. Welling. New York Public Library.

Papers of Richard W. G. Welling. New York Public Library.

Papers of Timothy S. Williams. New York Public Library.

Women's Municipal League of New York City Collection. New York Historical Society.

Documents

New York State Assembly. "Sixth Annual Report of the Factory Inspectors of the State of New York," Assembly Documents, No. 62, 115th Sess. Albany: J. B. Lyon, 1892.

________. "Tenth Annual Report of the Bureau of Statistics of Labor," Assembly Documents, No. 43, 116th Sess. Albany: J. B. Lyon, 1893.

________. "Seventh Annual Report of the Factory Inspectors of the State of New York," Assembly Documents, No. 63, 116th Sess. Albany: J. B. Lyon, 1893.

_______. Journal. 120th, 122nd, 130th Sessions. Albany: J. B. Lyon, 1897, 1899, 1907.

New York State Senate. "Ninth Annual Report of the Board of Railroad Commissioners on the Railroads of the State," Senate Documents, No. 6, 115th Sess. Albany: J. B. Lyon, 1892.

_______. "Communication from the Governor Relative to Certain Bills Now in His Hands Awaiting Executive Action," Senate Documents, No. 38, 115th Sess. Albany: J. B. Lyon, 1892.

_______. "Report of the Committee on Claims," Senate Documents, No. 40, 115th Sess. Albany: J. B. Lyon, 1892.

_______. "Communication from the Board of Estimate and Apportionment of the City of New York," Senate Documents, No. 28, 116th Sess. Albany: J. B. Lyon, 1893.

_______. "Report of the Special Committee Appointed To Investigate the Police Department of the City of New York," Senate Documents, No. 25, 118th Sess. Albany: J. B. Lyon, 1895.

New York State Assembly. "Minority Report of the Special Committee Appointed To Investigate the Police Department of the City of New York," Senate Documents, No. 29, 118th Sess. Albany: J. B. Lyon, 1895.

_______. "Annual Message of the Governor of the State of New York," Senate Documents, No. 2, 121st Sess. Albany: J. B. Lyon, 1898.

_______. "First Annual Report of the Commission of Gas and Electricity of the State of New York, Senate Documents, No. 15, 129th Sess. Albany: J. B. Lyon, 1906.

_______. "Report of the Public Service Commission, First District, State of New York," Senate Documents, No. 20, 131st Sess. Albany: J. B. Lyon, 1908.

_______. Journal. 115th, 120th, 121st Sessions. Albany: J. B. Lyon, 1892, 1897, 1898.

State of New York. Annual Report of the Public Service Commission, First District, 1907-1916. Albany: J. B. Lyon, 1908-1917.

________. First Annual Report of the Public Service Commission, 1921. Albany: J. B. Lyon, 1922.

Proceedings of the Board of Aldermen of the City of New York, 1892-1897.

Public Papers of Charles Evans Hughes. 4 vols. Albany: J. B. Lyon, 1908-1911.

Pamphlets and Reports

Anderson, William. The Work of Public Service Commissions with Special Reference to the New York Commissions. Minneapolis: University of Minnesota, 1913.

Bauer, John. The Public Utility Franchise: Its Functions and Terms Under State Regulation. Chicago: Public Administration Service, 1946.

Brandt, Lilian. Growth and Development of AICP and COS. New York: Community Service Society, 1942.

Bruere, Henry. Administrative Reorganization and Constructive Work in the Government of the City of New York. New York: Bureau of City Chamberlain, 1915.

Bruere, Henry. The Reorganization of the Office of Chamberlain. New York: Clarence S. Nathan, Inc., 1914.

Bureau of Municipal Research. Municipal Reform Through Revision of Business Methods. New York: Bureau of Municipal Research, 1910.

Chamber of Commerce of the State of New York. Annual Report, 1890-1914. New York: Chamber of Commerce, 1891-1915.

________. Catalogue of Portraits. New York: Chamber of Commerce, 1924.

_____. Rapid Transit in New York City and in Other Great Cities. New York: Chamber of Commerce, 1905.

Charities Organization Society of New York City. Annual Report, 1892, 1896-1897, 1899-1908, 1911-1913, 1916-1917.

Children's Aid Society of the City of New York. Annual Report, 1897, 1900-1901, 1905, 1916.

Citizens' Union. Campaign Book of the Citizens' Union, 1897, 1902, 1903. New York: Committee on Press and Literature of the Citizens' Union [1897, 1902, 1903].

_____. Annual Report, 1897-1924.

_____. the Citizens' Union of the City of New York, 1897-1931, What It Is, What It Does. New York: Citizens' Union, 1931.

_____. Report of the Committee on the Board of Aldermen, 1911-1917.

_____. Report of the Committee on Legislation, 1906-1917.

_____. The Searchlight. Vols. I-XII (1911-1922).

_____. Seth Low for Mayor: His Life and Public Record. New York: Committee on Press and Literature of the Citizens' Union [n.d.].

City Club of New York. Annual Report, 1894-1916.

_____. Some of the Things the City Club Has Done, 1892-1902. [n.d.]

City Reform Club of New York. Record of New York City Senators and Assemblymen in the State Legislature, 1892-1893.

Merchants' Association of New York. An Inquiry into the Conditions Relating to the Water-Supply of the City of New York. New York: Merchants' Association, 1900.

_____. the Merchants' Association of New York: Synopsis of the Greater New York Charter. New York: Merchants' Association, 1921.

_____. Passenger Transportation Service in the City of New York. New York: Merchants' Association, 1903.

_____. Waste of Water in New York and Its Reduction by Meters and Inspection. New York: Merchants' Association, 1906.

_____. Yearbook, 1899-1915. Issues from 1899-1904 called President's Report.

National Municipal League. A Municipal Program: Report of a Committee of the National Municipal League. New York: Macmillan Co., 1900.

_____. Proceedings of the National Conference for Good City Government and the Annual Meeting of the National Municipal League, Vol. I. Philadelphia: National Municipal League, 1895.

Our City, The Story of a Progressive People. New York: Regular Democratic Organization of Greater New York, 1909.

Jacob A. Riis; A Sketch of His Life and Work. New York: Macmillan Co. [n.d.]

Winslow, C.-E.A. The Contribution of Hermann Biggs to Public Health. Reprinted from The American Review of Tuberculosis, Vol. XX, No. 1 (July, 1929) by the New York Tuberculosis and Health Association [n.d.].

Newspapers and Periodicals

American Review of Reviews, 1905-1917.
Harper's Weekly, 1893-1901, 1906.
Independent, 1904-1905.
Literary Digest, 1905-1906.
National Municipal Review, 1912-1918.
New York Evening World, 1910, 1913.
New York Times, 1905, 1907.
New York World, 1893-1914.
Outlook, 1905-1907, 1915-1919.
Real Estate Record and Guide, 1893.

Unpublished Works

Dubofsky, Melvyn. "New York City Labor in the Progressive Era, 1910-1918: A Study of Organized Labor in an Era of Reform." Ph.D. dissertation, University of Rochester, 1960.

Gaus, John M. "A Study of Research in Public Administration." Prepared for the Advisory Committee on Public Administration of the Social Science Research Council. New York, 1930.

Hochman, William R. "William J. Gaynor: The Years of Fruition." Ph.D. dissertation, Columbia University, 1955.

Knerr, George Francis. "The Mayoral Administration of William L. Strong, New York City, 1895-1897." Ph.D. dissertation, New York University, 1957.

Articles in Periodicals

Allen, William H. "'The Business of Citizenship' in New York City," American Review of Reviews, XL (November, 1909), 594-601.

________. "Jeopardizing Reform in New York City," National Municipal Review, V (October, 1916), 656-658.

________. "New York's First Budget Exhibit," American Review of Reviews, XXXVIII (December, 1908), 686-688.

________. "Training Men and Women for Public Service," Annals, XLI (May, 1912), 307-312.

________. "Tuberose Obstacles to Reform in New York City," National Municipal Review, V (July, 1916), 419-426.

Anderson, William G. "Progressivism: An Historiographical Essay," The History Teacher, VI (May, 1973), 427-452.

Arndt, Walter T. "A Quarter Century of the Citizens Union: A Chapter in the Political History of New York City," Searchlight, XII (December, 1922), 1-22.

________. "The Defeated New York Constitution," National Municipal Review, V (January, 1916), 92-101.

Baker, Paula. "The Domestication of Politics: Women and American Political Society, 1780-1920," American Historical Review, LXXXIX (June, 1984), 620-647.

Baker, Ray Stannard. "The Subway 'Deal,'" McClure's, XXIV (March, 1905), 451-469.

Beard, Charles A. "The New York Constitutional Convention," National Municipal Review, IV (October, 1915), 637-645.

Bemis, Edward W. "Some Present Day Issues of Public Utility Regulation," Annals, LVII (January, 1915), 62-71.

Berrol, Selma. "When Uptown Met Downtown: Julia Richmond's Work in the Jewish Community of New York, 1880-1912," American Jewish History, LXX (1980), 35-51.

Birch, Eugenie Ladner, and Gardner, Deborah S. "The Seven-Percent Solution: A Review of Philanthropic Housing, 1870-1910," Journal of Urban History, VII (August, 1981), 408-438.

Braddock, J. Harold. "Efficiency Value of the Budget Exhibit," Annals, XLI (May, 1912), 151-157.

Brinkley, Allen. "Richard Hofstadter's The Age of Reform: A Reconsideration," Reviews in American History, XIII (September, 1985), 462-480.

Brown, M. Craig, and Halaby, Charles N. "Machine Politics in America, 1870-1945," Journal of Interdisciplinary History, XVII (Winter, 1987), 587-612.

Bruere, Henry. "The Budget as an Administrative Program," Annals, LXII (November, 1915), 176-191.

________. "Development of Standards in Municipal Government," Annals, LXI (September, 1915), 199-207.

________. "Efficiency in City Government," Annals, XLI (May, 1912), 1-22.

________. "Mayor Mitchel's Administration of the City of New York," National Municipal Review, V (January, 1916), 24-37.

_______. "New York's Municipal Problems," American Review of Reviews, XLVIII (October, 1913), 465-471.

_______. "Public Utilities Regulation in New York," Annals, XXXI (May, 1908), 1-17.

Bureau of Municipal Research. "The Accounting and Reporting Methods of the State of New York," Municipal Research, No. 74 (June, 1916).

_______. "Budget Legislation in Two States," Municipal Research, No. 70 (February, 1916).

_______. "Budget Systems," Municipal Research, No. 62 (June, 1915).

_______. "Citizen Agencies for Research in Government," Municipal Research, No. 77 (September, 1916).

_______. "The History of Appropriations in New York State," Municipal Research, No. 72 (April, 1916).

_______. "A National Program To Improve Methods of Government," Municipal Research, No. 71 (March, 1916).

_______. "The New York City Budget," Municipal Research, No. 88 (August, 1917).

_______. "The New York State Legislative Budget for 1917," Municipal Research, No. 86 (June, 1917).

_______. "Next Step in the Development of a Budget Procedure for the City of Greater New York," Municipal Research, No. 57 (January, 1915).

_______. "Responsible Government," Municipal Research, No. 69 (January, 1916).

_______. "Some Results and Limitations of Central Financial Control," Municipal Research, No. 81 (January, 1917).

_______. "Standardization of Public Employments," Municipal Research, No. 76 (August, 1916).

_______. "A State Budget," Municipal Research, No. 58 (February, 1915).

_______. "The State Movement for Efficiency and Economy," Municipal Research, No. 90 (October, 1917).

________. "Three Proposed Constitutional Amendments for Control of the Purse," Municipal Research, No. 73 (May, 1916).

Burks, J. D. "Efficiency Standards in Municipal Management," National Municipal Review, I (July, 1912), 364-371.

Cleveland, Frederick A. "The Need for Coordinating Municipal, State and National Activities," Annals, XLI (May, 1912), 23-39.

Clowes, J. H. "Efficiency in Water Revenue Collection," Annals, XLI (May, 1912), 86-92.

Clubb, Jerome M., and Allen, Howard W. "Collective Biography and the Progressive Movement: The 'Status Revolution' Revisited," Social Science History, I (Summer, 1977), 518-534.

Conway, Jill. "Women Reformers and American Culture, 1870-1930," Journal of Social History, Winter, 1971-72, pp. 164-175.

Cooper, Robert M. "Techniques of Public Control--An Appraisal of Methods," Annals, CCI (January, 1939), 1-16.

Daunton, Martin J. "Cities of Homes and Cities of Tenements: British and American Comparisons, 1870-1914," Journal of Urban History, XIV (May, 1988), 283-319.

Davis, Allen F. "Settlement Workers in Politics, 1890-1914," Review of Politics, XXVI (October, 1964), 505-517.

Dearstyne, Bruce W. "Regulation in the Progressive Era: The New York Public Service Commission," New York History, LVII (July, 1977), 331-347.

DeForest, Robert W. "Recent Progress in Tenement-House Reform," Annals, XXIII (March, 1904), 297-310.

Dorsett, Lyle W. "The City Boss and the Reformer: A Reappraisal," Pacific Northwest Quarterly, LXIII (October, 1972), 150-154.

Dougherty, J. Hampden. "Seth Low's Service in Behalf of Non-Partisan City Government," National Municipal Review, VI (March, 1917), 210-216.

Driscoll, Clement J. "The New York Police Investigation," National Municipal Review, II (April, 1913), 279-283.

Dubofsky, Melvyn. "Organized Labor and the Immigrant in New York City, 1900-1918," Labor History, II (Spring 1961), 182-201.

Dupree, A. Hunter, and Fishel, Leslie H., Jr. (eds.). "An Eyewitness Account of the New York Draft Riots, July, 1863," Mississippi Valley Historical Review, XLVII (December, 1960), 472-479.

Ebner, Michael H. "Urban History: Retrospect and Prospect," The Journal of American History, LXVIII (June, 1981) 69-84.

Epstein, Amy Kallman. "Multifamily Dwellings and the Search for Respectability: Origins of the New York Apartment House," Urbanism Past and Present, V (Summer, 1980), 29-39.

Eutsler, Roland B. "Public and Private Ownership of Water Supply Utilities," Annals, CCI (January, 1939), 89-95.

Felt, Jeremy P. "Vice Reform as a Political Technique: The Committee of Fifteen in New York, 1900-1901," New York History, LIV, (January, 1973), 24-51.

Filene, Peter G. "An Obituary for 'The Progressive Movement,'" American Quarterly, XXII (1970), 20-34.

Fisher, Edmund D. "Efficiency in the Fiscal Operations of Cities," Annals, XLI (May, 1912), 71-77.

Fitch, Charles E. "New York's Constitutional Convention," Harper's Weekly, XXXVIII (April 14, 1894), 347.

Foster, Mark. "City Planners and Urban Transportation: The American Response, 1900-1940," Journal of Urban History, V (May, 1979), 365-396.

Frisch, Michael H. "Urban Theorists, Urban Reform, and American Political Culture in the Progressive Period," Political Science Quarterly, XCVII (Summer, 1982), 295-315.

Gilbertson, H. S. "Municipal Revolution Under Mayor Mitchel," American Review of Reviews, LVI (September, 1917), 300-303.

Goodrich, E. P. "Efficiency in Highway Administration with Special Reference to Pavements," Annals, XLI (May, 1912), 115-126.

Grantham, Dewey, Jr. "The Progressive Era and the Reform Tradition," Mid-America, XLVI (October, 1964), 227-251.

Hammack, David C. "Problems in the Historical Study of Power in the Cities and Towns of the United States, 1800-1960," The American Historical Review, LXXXIII (April, 1978), 323-349.

Hays, Samuel P. "The Changing Political Structure of the City in Industrial America," Journal of Urban History, I (November, 1974), 6-38.

________. "The Politics of Reform in Municipal Government in the Progressive Era," Pacific Northwest Quarterly, LV (October, 1964), 157-169.

________. "The Social Analysis of American Political History, 1880-1920," Political Science Quarterly, LXXX (September, 1965), 373-394.

Hendrick, Burton J. "Governor Hughes," McClure's, XXX, Part I (March, 1908), 521-536 and Part II (April, 1908), 670-681.

Hobson, Wayne K. "Professionals, Progressives and Bureaucratization: A Reassessment," The Historian, XXXIX (August, 1977), 639-658.

Hopkins, George B. "The New York Bureau of Municipal Research," Annals, XLI (May, 1912), 235-244.

Huthmacher, J. Joseph. "Charles Evans Hughes and Charles Francis Murphy: The Metamorphosis of Progressivism," New York History, XLVI (January, 1965), 25-40.

Ihlder, John. "Old Dwellings for New," National Municipal Review, VI (November, 1917), 688-692.

Ingersoll, Raymond V. "An Estimate of Mayor Gaynor," National Municipal Review, III (January, 1914), 78-83.

_______. "The Recent New York City Fusion Campaign," National Municipal Review, VII (March, 1918), 187-189.

Johnston, Charles. "Charles Evans Hughes and the Contest in New York State," North American Review, CLXXXIII (November 2, 1906), 899-905.

Jones, Stiles P. "State Versus Local Regulation," Annals, LIII (May, 1914), 94-107.

Kaplan, Barry J. "Andrew H. Green and the Creation of a Planning Rationale: The Formation of Greater New York City, 1865-1890," Urbanism Past and Present, VIII (Summer, 1979), 32-41.

Kennedy, David M. "Overview: The Progressive Era," The Historian, XXXVII (May, 1975), 453-468.

Kennedy, John S. "The New York Public Service Commissions," Forum, XLVIII (November, 1912), 584-595.

Kerr, William Dunton. "Qualifications Needed for Public Utility Commissioners," Annals, XLII (May, 1914), 19-35.

Kirschner, Don S. "The Ambiguous Legacy: Social Justice and Social Control in the Progressive Era," Historical Reflections, II (Summer, 1975), 69-88.

Klein, Otto H. "Securing Efficiency Through a Standard Testing Laboratory," Annals, XLI (May, 1912), 93-102.

Lane, James B. "Jacob A. Riis and Scientific Philanthropy during the Progressive Era," Social Service Review, XLVII (March, 1973), 32-48.

Leonhauser, U. L. "A National Fund for Promoting Efficient Municipal Accounting and Reporting," Annals, XLI (May, 1912), 304-306.

Lubove, Roy. "The New York Association for Improving the Conditions of the Poor: The Formative Years," New York Historical Society Quarterly, XLIII (July, 1959), 307-327.

_______. "The Twentieth Century City: The Progressive as Municipal Reformer," Mid-America, XLI (October, 1959), 195-209.

Maltbie, Milo R. "The Fruits of Public Regulation in New York," Annals, XXXVII (January, 1911), 170-190.

Margulies, Herbert F. "Recent Opinion on the Decline of the Progressive Movement," Mid-America, XLV (October, 1963), 250-268.

Matthews, Franklin. "The Cost of Tammany Hall in Flesh and Blood," Harper's Weekly, XLIV (October 13, 1900), 957-960.

_______. "A Matter of Municipal Ownership and What Came of It," Harper's Weekly, XLI (October 30, 1897), 1074.

Mayers, Lewis. "The New York School Inquiry," National Municipal Review, III (April, 1914), 327-339.

McCormick, Richard L. "The Discovery that Business Corrupts Politics: A Reappraisal of the Origins of Progressivism," American Historical Review, LXXXVI (April, 1981), 247-274.

McCraw, Thomas K. "Regulation in America: A Review Article," Business History Review, XLIX (Summer, 1975), 159-183.

McDonald, Terrence J. "The Problem of the Political in Recent American Urban History: Liberal Pluralism and the Rise of Functionalism," Social History, X (1985), 323-345.

Monroe, Robert Grier. "The Gas, Electric Light, Water and Street Railway Services in New York City," Annals, XXVII (1906), 111-119.

Mosher, William E. "Defects of State Regulation of Public Utilities in the United States," Annals, CCI (January, 1939), 105-110.

Murphy, John J. "Franchise Grants in New York City," Annals, XXXI (May, 1908), 78-84.

________. "Non-Partisanship in Municipal Affairs as Illustrated by New York Experience," National Municipal Review, VI (March, 1917), 217-218.

N[elson], H[enry] L. "Tammany on Trial--The Mazet Investigation," Harper's Weekly, XLIII (September 9, 1899), 881-882.

Nord, David. "The Experts Versus the Experts: Conflicting Philosophies of Municipal Utility Regulation in the Progressive Era," Wisconsin Magazine of History, LVIII (Spring, 1975), 219-236.

Osofsky, Gilbert. "Progressivism and the Negro: New York, 1900-1915," American Quarterly, XVI (Summer, 1964), 153-168.

Outerbridge, E. H. "Lower Telephone Rates for New York City," Annals, LIII (May, 1914), 66-70.

Pease, Otis A. "Urban Reformers in the Progressive Era: A Reassessment," Pacific Northwest Quarterly, LXII (April, 1971), 49-58.

Pollock, Horatio M. "The Public Service Commission of the State of New York," Annals, XXXI (May, 1908), 649-658.

Prendergast, William A. "Efficiency Through Accounting," Annals, XLI (May, 1912), 43-56.

________. "New York City Finances," National Municipal Review, II (April, 1913), 221-229.

________. "Why New York City Needs a New School Plan," American Review of Reviews, LII (November, 1915), 584-588.

Price, Joseph M. "Fusion Mistakes and a Way Out," National Municipal Review, VII (March, 1918), 183-186.

Pultz, J. Leggett. "Economy and Efficiency in the Department of Water Supply, Gas and Electricity of New York City," Annals, XLI (May, 1912), 178-85.

Rightor, C. E. "Recent Progress in Municipal Budgets and Accounts," National Municipal Review, V (July, 1916), 403-410.

Riis, Jacob A. "The New York Tenement-House Commission," American Monthly Review of Reviews, XXI (June, 1900), 689-696.

________. "Small Parks and Public-School Play-Grounds," Harper's Weekly, XLI (September 11, 1897), 903.

________. "The Tenement House Exhibition," Harper's Weekly, XLIV (February 3, 1900), 104.

________. "The Tenement-House Problem," Harper's Weekly, XXXIX (January 12, 1895), 42-43.

Ritchie, Donald A. "The Gary Committee: Businessmen, Progressives, and Unemployment in New York City, 1914-1915," The New York Historical Society Quarterly, LVII (October, 1973), 327-347.

Rodgers, Daniel T. "In Search of Progressivism," Reviews in American History, X (December, 1982), 113-132.

Sait, Edward M. "Research and Reference Bureaus," National Municipal Review, II (January, 1913), 48-56.

Sands, Herbert R., and Lindars, Fred W. "Efficiency in Budget Making," Annals, XLI (May, 1912), 138-150.

Schurz, Carl. "Bossism in New York," Harper's Weekly, XLI (November 13, 1897), 1119.

________. "The Municipal Situation," Harper's Weekly, XLI (June 5, 1897), 555.

________. "The Progress of Independence in Municipal Elections," Harper's Weekly, XLI (October 16, 1897), 1023.

________. "The Right To Nominate," Harper's Weekly, XLI (July 3, 1897), 651.

________. "The Task of the Citizens' Union," Harper's Weekly, XLI (December 11, 1897), 1215.

________. "True Non-Partisanship," Harper's Weekly, XLI (October 2, 1897), 967.

Scott, Anne F. "On Seeing and Not Seeing: A Case of Historical Invisibility," Journal of American History, LXXI (June, 1984), 13-17.

Sharfman, I. Leo. "Commission Regulation of Public Utilities: A Survey of Legislation," Annals, LIII (May, 1914), 1-18.

S[haw], A[lbert]. "New York City's Government by Experts," American Review of Reviews, XLIX (February, 1914), 171-178.

Shaw, William B. "The Men Who Are Governing New York City," American Review of Reviews, XLI (March, 1910), 300-307.

Sklar, Kathryn Kish. "Hull House in the 1980s: A Community of Women Reformers," Signs, X (Summer, 1985), 658-677.

Skolnik, Richard. "Civic Group Progressivism in New York City," New York History, LI (July, 1970), 411-439.

Smith, J. Allen. "Effects of State Regulation of Public Utilities Upon Municipal Home Rule," Annals, LIII (May, 1914), 85-93.

Stave, Bruce M., et al. "A Reassessment of the Urban Political Boss: An Exchange of Views," The History Teacher, XXI (May, 1988), 293-312.

Stowe, Lyman Beecher. "How New York Deals with Her Public Service Companies," American Review of Reviews, XLII (August, 1910), 211-221.

Swan, Herbert S. "How Zoning Works in New York," National Municipal Review, VII (May, 1918), 244-254.

Swett, Steven E. "The Test of a Reformer, A Study of Seth Low, New York City Mayor, 1902-1903," New York Historical Society Quarterly, XLIV (January, 1960), 5-41.

Tanzer, Lawrence A. "Legislative Interference in Municipal Affairs and the Home Rule Program in New York," National Municipal Review, II (October, 1913), 597-604.

Taylor, George R. "The Beginnings of Mass Transportation in Urban America, Part I," Smithsonian Journal of History, II (Summer, 1966), 35-50.

Teaford, Jon C. "Finis for Tweed and Steffens: Rewriting the History of Urban Rule," Reviews in American History, X (December, 1982), 133-149.

Thelen, David P. "Social Tensions and the Origins of Progressivism," The Journal of American History, LVI (September, 1969), 323-341.

Tobin, Eugene M. "The Progressive as Humanitarian: Jersey City's Search for Social Justice, 1890-1917," New Jersey History, XCIII (1975), 78-98.

Urban, Wayne. "Organized Teachers and Educational Reform During the Progressive Era, 1890-1920," History of Education Quarterly, XVI (1976), 35-52.

Warner, John DeWitt. "The 'Sweating System' in New York City," Harper's Weekly, XXXIX (February 9, 1895), 135.

Weinstein, James. "Organized Business and the City Commission and Manager Movements," Journal of Southern History, XXVIII (May, 1962), 166-182.

Welton, Benjamin F. "The Problem of Securing Efficiency in Municipal Labor," Annals, XLI (May, 1912), 103-114.

Wesser, Robert F. "Charles Evans Hughes and the Urban Sources of Political Progressivism," New York Historical Society Quarterly, (October, 1966), 365-400.

________. "Theodore Roosevelt: Reform and Reorganization of the Republican Party in New York, 1901-1906," New York History, XLVI (July, 1965), 230-252.

Whitten, Robert H. "The Building Zone Plan of New York City," National Municipal Review, VI (May, 1917), 325-336.

Wilcox, Delos F. "The Crisis in Public Service Regulation in New York," National Municipal Review, IV (October, 1915), 547-563.

________. "Effects of State Regulation Upon the Municipal Ownership Movement," Annals, LIII (May, 1914), 71-84.

________. "Experts, Ethics and Public Policy in Public Utilities," National Municipal Review, VI (July, 1917), 472-485.

________. "The New York Subway Contracts," National Municipal Review, II (July, 1913), 375-391.

Williams, Marilyn Thornton. "New York City's Public Baths: A Case Study in Urban Progressive Reform," Journal of Urban History, VII (November, 1980), 49-81.

Wright, Henry C. "Development of Transit Control in New York City," Annals, XXXI (May, 1908), 552-575.

Wyman, Roger E. "Middle-Class Voters and Progressive Reform: The Conflict of Class and Culture," The American Political Science Review, LXVIII (June, 1974), 488-504.

Books

Alexander, DeAlva S. Four Famous New Yorkers. Vol. IV of Political History of the State of New York 1882-1905. New York: Henry Holt and Co., 1923.

Allen, William H. Civics and Health. Boston: Ginn and Company, 1909.

Allswang, John M. Bosses, Machines, and Urban Voters. Port Washington, NY: Kennikat Press, 1977.

Arent, Leonora. Electric Franchises in New York City. New York: Columbia University, 1919.

Banfield, Edward C., and Wilson, James Q. City Politics. Cambridge: Harvard University Press, 1963.

Baritz, Loren. The Servants of Power: A History of the Use of Social Science in American Industry. Middletown, Conn.: Wesleyan University Press, 1960.

Bender, Thomas. Community and Social Change in America. Baltimore: Johns Hopkins University Press, 1978.

Boulding, Kenneth E. The Organizational Revolution. New York: Harper and Bros., 1953.

Boyer, Paul. Urban Masses and Moral Order in America, 1820-1920. Cambridge, Mass.: Harvard University Press, 1978.

Bradley, Joseph F. The Role of Trade Associations and Professional Business Societies in America. University Park, Penn.: Pennsylvania State University Press, 1965.

Bremner, Robert H. American Philanthropy. Chicago: University of Chicago Press, 1960.

Bruere, Henry. The New City Government. New York: D. Appleton and Co., 1912.

Buenker, John D. Urban Liberalism and Progressive Reform. New York: Charles Scribner's Sons, 1973.

________, Burnham, John C., and Crunden, Robert M. Progressivism. Cambridge, Mass.: Schenkman Publishing Company, Inc., 1977.

Carman, Harry J. The Street Surface Railway Franchises of New York City. New York: Columbia University, 1919.

Cavallo, Dominick. Muscles and Morals: Organized Playgrounds and Urban Reform, 1880-1920. Philadelphia: University of Pennsylvania Press, 1981.

Cheape, Charles W. Moving the Masses: Urban Public Transit in New York, Boston, and Philadelphia, 1880-1912. Cambridge, Mass.: Harvard University Press, 1980.

Chessman, G. Wallace. Governor Theodore Roosevelt: The Albany Apprenticeship, 1898-1900. Cambridge: Harvard University Press, 1965.

Childs, Richard S. Civic Victories, The Story of An Unfinished Revolution. New York: Harper and Bros., 1952.

Cleveland, Frederick A. Chapters on Municipal Administration and Accounting. New York: Longmans, Green, and Co., 1909.

________. Funds and Their Uses. New York: D. Appleton and Co., 1902.

________. Organized Democracy. New York: Longmans, Green, and Co., 1913.

________, and Buck, Arthur Eugene. The Budget and Responsible Government. New York: The Macmillan Co., 1920.

Cohen, Sol. Progressives and Urban School Reform: The Public Education Association of New York City 1895-1954. New York: Columbia University, 1964.

Colburn, David R., and Pozzetta, George E. (eds.). Reform and Reformers in the Progressive Era. Westport, Conn.: Greenwood Press, 1983.

Connelly, Mark Thomas. The Response to Prostitution in the Progressive Era. Chapel Hill: University of North Carolina Press, 1980.

Crunden, Robert M. Ministers of Reform: The Progressives' Achievement in American Civilization, 1889-1920. Urbana: University of Illinois Press, 1984.

Dahlberg, Jane S. The New York Bureau of Municipal Research, Pioneer in Government Administration. New York: New York University Press, 1966.

Danbom, David B. "The World of Hope": Progressives and the Struggle for an Ethical Public Life. Philadelphia: Temple University Press, 1987.

Davis, Allen F. Spearheads for Reform: The Social Settlements and the Progressive Movement, 1890-1914. New York: Oxford University Press, 1967.

Davis, Keith, and Blomstrom, Robert L. Business and Its Environment. New York: McGraw-Hill Book Co., 1966.

DeForest, Robert W., and Veiller, Lawrence (eds.). The Tenement House Problem. 2 vols. New York: The Macmillan Co., 1903.

Devine, Edward T. Efficiency and Relief; A Programme of Social Work. New York: Columbia University Press, 1906.

________. When Social Work Was Young. New York: The Macmillan Co., 1939.

Devlin, Thomas C. Municipal Reform in the United States. New York: G. P. Putnam's Sons, The Knickerbocker Press, 1896.

Dubofsky, Melvyn. Industrialism and the American Worker, 1865-1920. 2nd ed. Arlington Heights, Ill.: Harlan Davidson, Inc., 1985.

________. When Workers Organize: New York City in the Progressive Era. Amherst: University of Massachusetts Press, 1968.

Duffy, John. A History of Public Health in New York City, 1866-1966. New York: Russell Sage Foundation, 1974.

East, John Porter. Council-Manager Government; The Political Thought of Its Founder, Richard S. Childs. Chapel Hill: University of North Carolina Press, 1965.

Ebner, Michael H., and Tobin, Eugene M. (eds.). The Age of Urban Reform: New Perspectives on the Progressive Era. Port Washington, N.Y.: Kennikat Press, 1977.

Ellis, David M. et al. A Short History of New York State. Ithaca, N.Y.: Cornell University Press, 1957.

Etzioni, Amitai. Modern Organizations. Englewood Cliffs, N.J.: Prentice-Hall, Inc., 1964.

Faulkner, Harold U. Politics, Reform, and Expansion. New York: Harper and Bros., 1959.

Felt, Jeremy P. Hostages of Fortune; Child Labor Reform in New York State. Syracuse, N.Y.: Syracuse University Press, 1965.

Fitzpatrick, Edward A. (ed.). Experts in City Government. New York: D. Appleton and Co., 1919.

Fleming, Donald. William H. Welch and the Rise of Modern Medicine. Boston: Little, Brown and Co., 1954.

Flick, Alexander (ed.). History of the State of New York. 10 vols. New York: Columbia University Press, 1933-1937.

Fosdick, Raymond B. Chronicle of a Generation. New York: Harper and Bros., 1958.

Fox, Kenneth. Better City Government: Innovation in American Urban Politics, 1850-1937. Philadelphia: Temple University Press, 1977.

Galishoff, Stuart. Safeguarding the Public Health: Newark, 1895-1918. Westport, Conn.: Greenwood Press, 1975.

Garrett, Charles. The LaGuardia Years, Machine and Reform Politics in New York City. New Brunswick, N.J.: Rutgers University Press, 1961.

Gaus, John M. Reflections on Public Administration. University, Alabama: University of Alabama Press, 1947.

Gill, Norman. Municipal Research Bureaus. Washington, D.C.: American Council on Public Affairs, 1944.

Gillette, Howard, Jr., and Miller, Zane L. (eds.). American Urbanism: A Historiographical Review. New York: Greenwood Press, 1987.

Gosnell, Harold F. Boss Platt and His New York Machine. Chicago: University of Chicago Press, 1924.

Grantham, Dewey W. Southern Progressivism: The Reconciliation of Progress and Tradition. Knoxville, University of Tennessee Press, 1983.

Green, Constance M. The Rise of Urban America. New York: Harper and Row Publishers, 1965.

Greer, Thomas H. American Social Reform Movements: Their Pattern Since 1865. New York: Prentice-Hall, Inc., 1949.

Griffith, Ernest S. A History of American City Government: The Conspicuous Failure, 1870-1900. Washington, D.C.: University Press of America, 1983.

________. A History of American City Government: The Progressive Years and Their Aftermath, 1900-1920. Washington, D.C.: University Press of America, 1983.

Gould, Lewis L. (ed.). The Progressive Era. Syracuse: Syracuse University Press, 1974.

Gulick, Luther. The National Institute of Public Administration: A Progress Report. New York: National Institute of Public Administration, 1928.

Haber, Samuel. Efficiency and Uplift: Scientific Management in the Progressive Era, 1890-1920. Chicago: University of Chicago Press, 1964.

Hammack, David C. Power and Society: Greater New York at the Turn of the Century. New York: Columbia University Press, 1987.

Handlin, Oscar. Al Smith and His America. Boston: Little Brown and Co., 1958.

Hapgood, Norman, and Moskowitz, Henry. Up From the City Streets: Alfred E. Smith. New York: Harcourt, Brace and Co., 1927.

Harbaugh, William H. Power and Responsibility: The Life and Times of Theodore Roosevelt. New York: Farrar, Straus and Cudahy, 1961.

Harris, Carl V. Political Power in Birmingham, 1871-1921. Knoxville: University of Tennessee Press, 1977.

Hays, Samuel P. American Political History as Social Analysis. Knoxville: University of Tennessee Press, 1980.

________. Conservation and the Gospel of Efficiency: The Progressive Conservation Movement, 1890-1920. Cambridge, Mass.: Harvard University Press, 1959.

________. The Response to Industrialism: 1885-1914. Chicago: University of Chicago Press, 1957.

Henderson, Thomas M. Tammany Hall and the New Immigrants: The Progressive Years. New York: Arno Press, 1976.

Hofstadter, Richard. The Age of Reform. New York: Alfred A. Knopf, Inc., 1955.

Holli, Melvin G. Reform in Detroit: Hazen S. Pingree and Urban Politics. New York: Oxford University Press, 1969.

Howe, Frederic C. The City: The Hope of Democracy. New York: Charles Scribner's Sons, 1905.

________. The Modern City and Its Problems. New York: Charles Scribner's Sons, 1915.

Howe, M. A. DeWolfe. John Jay Chapman and His Letters. Boston: Houghton Mifflin Co., The Riverside Press, Cambridge, 1937.

Howe, Wirt. New York at the Turn of the Century, 1899-1916. Toronto, Canada: [privately printed], 1946.

Jackson, Anthony. A Place Called Home: A History of Low-Cost Housing in Manhattan. Cambridge, Mass.: MIT Press, 1976.

Jaher, Frederic Cople. The Urban Establishment: Upper Strata in Boston, New York, Charleston, Chicago, and Los Angeles. Urbana: University of Illinois Press, 1982.

Karl, Barry D. Executive Reorganization and Reform in the New Deal: The Genesis of Administrative Management, 1900-1939. Cambridge, Mass.: Harvard University Press, 1963.

King, Clyde L. (ed.). The Regulation of Municipal Utilities. New York: D. Appleton and Co., 1914.

Klein, Henry H. Politics, Government and the Public Utilities in New York City. New York: Isaac Goldmann Co., 1933.

Kolko, Gabriel. The Triumph of Conservatism: A Reinterpretation of American History, 1900-1916. New York: The Free Press, 1963.

Kurland, Gerald. Seth Low: The Reformer in an Urban and Industrial Age. New York: Twayne Publishers, Inc., 1971.

Leonard, John W. History of the City of New York, 1609-1909. 2 vols. New York: Journal of Commerce and Commercial Bulletin, 1910.

Leopold, Richard W. Elihu Root and the Conservation Tradition. Boston: Little, Brown and Co., 1954.

Lewinson, Edwin R. John Purroy Mitchel, The Boy Mayor of New York. New York: Astra Books, 1965.

Lewis, Alfred H. The Boss, and How He Came To Rule New York. New York: A. S. Barnes and Co., 1904.

Link, Arthur S., and McCormick, Richard L. Progressivism. Arlington Heights, Ill.: Harlan Davidson, Inc., 1983.

Low, Benjamin R. C. Seth Low. New York: C. P. Putnam's Sons, The Knickerbocker Press, 1925.

Lowi, Theodore J. At the Pleasure of the Mayor: Patronage and Power in New York City, 1898-1958. London: Free Press of Glencoe, 1964.

Lubove, Roy. The Professional Altruist: The Emergence of Social Work as a Career 1880-1930. Cambridge, Mass.: Harvard University Press, 1905.

________. The Progressives and the Slums: Tenement House Reform in New York City, 1890-1917. Pittsburgh: University of Pittsburgh Press, 1962.

Makielski, S. J., Jr. The Politics of Zoning; The New York Experience. New York: Columbia University Press, 1966.

Malone, Dumas (ed.). Dictionary of American Biography. Vols. XX, IX. New York: Charles Scribner's Sons, 1936.

Mann, Arthur. LaGuardia, A Fighter Against His Times, 1882-1933. New York: J. B. Lippincott Co., 1959.

McCormick, Richard L. From Realignment to Reform: Political Change in New York State, 1893-1910. Ithaca: Cornell University Press, 1979.

________. The Party Period and Public Policy: American Politics from the Age of Jackson to the Progressive Era. New York: Oxford University Press, 1986.

McDonald, Terrence J. The Parameters of Urban Fiscal Policy: Socioeconomic Change and Political Culture in San Francisco, 1860-1906. Berkeley: University of California Press, 1986.

________, and Ward, Sally K. The Politics of Urban Fiscal Policy. Beverly Hills: Sage Publications, 1984.

Miller, Zane L. Boss Cox's Cincinnati: Urban Politics in the Progressive Era. New York: Oxford University Press, 1968.

Milne, Gordon. George William Curtis and the Genteel Tradition. Bloomington: Indiana University Press, 1956.

Mohl, Raymond A. The New City: Urban America in the Industrial Age, 1860-1920. Arlington Heights, Ill.: Harlan Davidson, Inc., 1985.

Morgan, H. Wayne (ed.). The Gilded Age, A Reappraisal. New York: Syracuse University Press, 1963.

Mosher, William E., and Crawford, Finla G. Public Utility Regulation. New York: Harper and Bros., 1933.

Mowry, George E. The California Progressives. Chicago: Quadrangle Books, 1951.

________. The Era of Theodore Roosevelt, 1900-1912. New York: Harper and Row Publishers, 1958.

Munro, William Bennett. Principles and Methods of Municipal Administration. New York: The Macmillan Co., 1920.

Nevins, Allan. Abram S. Hewitt with Some Account of Peter Cooper. New York: Harper and Bros., 1935.

________, and Kraut, John A. (eds.). The Greater City New York, 1898-1948. New York: Columbia University Press, 1948.

Patton, Clifford W. The Battle for Municipal Reform, Mobilization and Attack, 1875-1900. Washington, D.C.: American Council on Public Affairs, 1940.

Peel, Roy V. The Political Clubs of New York City. New York: A. P. Putnam's Sons, 1935.

Perkins, Dexter. Charles Evans Hughes and American Democratic Statesmanship. Boston: Little, Brown and Co., 1956.

Perry, Elisabeth Israels. Belle Moskowitz: Feminine Politics and the Exercise of Power in the Age of Alfred E. Smith. New York: Oxford University Press, 1987.

Peterson, Lorin. The Day of the Mugwump. New York: Random House, 1961.

Philpott, Thomas Lee. The Slum and the Ghetto: Neighborhood Deterioration and Middle-Class Reform, Chicago, 1880-1930. New York: Oxford University Press, 1978.

Pivar, David J. Purity Crusade Sexual Morality and Social Control, 1868-1900. Westport, Conn.: Greenwood Press, Inc., 1973.

Pusey, Merlo J. Charles Evans Hughes. 2 vols. New York: The Macmillan Co., 1951.

Reed, Thomas Harrison. Municipal Government in the United States. Rev. ed. New York: D. Appleton-Century Co., 1934.

Rice, Bradley R. Progressive Cities: The Commission Government Movement in America, 1901-1920. Austin: University of Texas Press, 1977.

Riis, Jacob A. Children of the Tenements. New York: The Macmillan Co., 1903.

________. How the Other Half Lives; Studies Among the Tenements of New York. New York: Charles Scribner's Sons, 1899.

________. The Making of an American. Harper Torchbooks. Edited with an Introduction by Roy Lubove. New York: Harper and Row Publishers, 1966.

________. A Ten Years' War; An Account of the Battle with the Slum in New York. New York: Houghton, Mifflin and Company, 1900.

Riordan, William L. Plunkitt of Tammany Hall. New York: E. P. Dutton and Co., Inc., 1963.

Rischin, Moses. The Promised City; New York's Jews, 1870-1914. Cambridge, Mass.: Harvard University Press, 1962.

Rodgers, Cleveland, and Rankin, Rebecca B. New York: The World's Capital City. New York: Harper and Bros., 1948.

Rosen, George. A History of Public Health. New York: MD Publications, Inc., 1958.

Rosenzweig, Roy. Eight Hours for What We Will; Workers and Leisure in an Industrial City, 1870-1920. Cambridge: Cambridge University Press, 1983.

Ruggles, C. O. Aspects of the Organization, Functions, and Financing of State Public Utility Commissions. Cambridge, Mass.: Harvard University Graduate School of Business Administration, Bureau of Business Research, 1937.

Sayre, Wallace S., and Kaufman, Herbert. Governing New York City. New York: Russell Sage Foundation, 1960.

Schiesl, Martin J. The Politics of Efficiency: Municipal Administration and Reform in America, 1880-1920. Berkeley: University of California Press, 1977.

Schlesinger, Arthur M. The Rise of the City, 1878-1898. Vol. X of A History of American Life. New York: The Macmillan Co., 1933.

Schneider, David M., and Deutsch, Albert. The History of Public Welfare in New York State, 1867-1940. Chicago: University of Chicago Press, 1941.

Seidman, Harold. Investigating Municipal Administration. New York: Institute of Public Administration, Columbia University, 1941.

Seitz, Don C. Joseph Pulitzer, His Life and Letters. New York: Simon and Schuster, 1924.

Shaw, Frederick. The History of the New York City Legislature. New York: Columbia University Press, 1954.

Sheehan, Donald, and Syrett, Harold C. (eds.). Essays in American Historiography, Papers presented in Honor of Allan Nevins. New York: Columbia University Press, 1960.

Smith, Mortimer. William Jay Gaynor: Mayor of New York. Chicago: Henry Regnery Company, 1951.

Spurr, Henry C. Guiding Principles of Public Service Regulation. 2 vols. Rochester, N.Y.: Public Utilities Reports, Inc., 1924.

Stave, Bruce M. (ed.) Socialism and the Cities. Port Washington, N.Y.: Kennikat Press, 1975.

Steffens, Lincoln. The Autobiography of Lincoln Steffens. New York: Harcourt, Brace and Co., 1931.

Stewart, Frank Mann. A Half Century of Municipal Reform, the History of the National Municipal League. Berkeley: University of California Press, 1950.

________. The National Civil Service Reform League; History, Activities, and Problems. Austin: University of Texas, 1929.

Swanberg, W. A. Citizen Hearst: A Biography of William Randolph Hearst. New York: Charles Scribner's Sons, 1961.

Syrett, Harold C. The City of Brooklyn, 1865-1898; A Political History. New York: Columbia University Press, 1944.

________. (ed.). The Gentleman and the Tiger; The Autobiography of George B. McClellan, Jr. New York: J. B. Lippincott Co., 1956.

Teaford, Jon C. The Unheralded Triumph: City Government in America, 1870-1900. Baltimore: Johns Hopkins University Press, 1984.

Thelen, David P. The New Citizenship: Origins of Progressivism in Wisconsin, 1885-1900. Columbia: University of Missouri Press, 1972.

Tolman, William H., and Hull, William I. Handbook of Sociological Information with Especial Reference to New York City. New York: G. P. Putnam's Sons, Knickerbocker Press, 1894.

Trattner, Walter I. From Poor Law to Welfare State: A History of Social Welfare in America. 2nd ed. New York: The Free Press, 1979.

The Triumph of Reform: A History of the Great Political Revolution, November Sixth, 1894. New York: Souvenir Publishing Company, 1895.

Tyack, David B. The One Best System: A History of American Urban Education. Cambridge, Mass.: Harvard University Press, 1974.

Veiller, Lawrence. Housing Reform; A Hand-Book for Practical Use in American Cities. New York: Charities Publication Committee, 1911.

Wald, Lillian D. The House on Henry Street. New York: Henry Holt and Company, 1915.

_______. Windows on Henry Street. Boston: Little, Brown and Company, 1934.

Waldo, Dwight. The Administrative State; A Study of the Political Theory of American Public Administration. New York: The Ronald Press Company, 1948.

Walker, James B. Fifty Years of Rapid Transit, 1864-1917. New York: Law Printing Company, 1918.

Ware, Louise. Jacob A. Riis, Police Reporter, Reformer, Useful Citizen. New York: D. Appleton-Century Company, 1938.

Watson, Frank D. The Charity Organization Movement in the United States; A Study in American Philanthropy. New York: The Macmillan Company, 1922.

Weinstein, James. The Corporate Ideal in the Liberal State: 1900-1918. Boston: Beacon Press, 1968.

Weiss, Nancy J. Charles Francis Murphy, 1858-1924: Respectability and Responsibility in Tammany Politics. Northampton, Mass.: Smith College, 1968.

Welling, Richard. As the Twig Is Bent. New York: G. P. Putnam's Sons, 1942.

Werner, Morris R. Tammany Hall. Garden City, N.Y.: Doubleday, Doran and Company, Inc., 1928.

Wesser, Robert F. Charles Evans Hughes: Politics and Reform in New York, 1905-1910. Ithaca, N.Y.: Cornell University Press, 1967.

Wheeler, Everett P. Sixty Years of American Life. New York: E. P. Dutton and Company, 1917.

White, Leonard D. The Republican Era, 1869-1901. New York: The Macmillan Company, 1958.

Wiebe, Robert H. Businessmen and Reform: A Study of the Progressive Movement. Cambridge: Harvard University Press, 1962.

_______. The Search for Order, 1877-1920. New York: Hill and Wang, 1967.

Wilcox, Delos F. The American City: A Problem in Democracy. New York: The Macmillan Company, 1904.

_______. Municipal Franchises. 2 vols. Rochester, N.Y.: The Gervaise Press, 1910.

Winslow, C.-E.A. The Evolution and Significance of the Modern Public Health Campaign. New Haven: Yale University Press, 1923.

_______. The Life of Hermann M. Biggs, Physician and Statesman of the Public Health. Philadelphia: Lea and Febiger, 1929.

Woodward, C. Vann (ed.). The Comparative Approach to American History. New York: Basic Books, Inc., Publishers, 1968.

Yellowitz, Irwin. Labor and the Progressive Movement in New York State, 1897-1916. Ithaca, N.Y.: Cornell University Press, 1965.

Zeller, Belle. Pressure Politics in New York; A Study of Group Representation Before the Legislature. New York: Prentice-Hall, Inc., 1937.

Zueblin, Charles. American Municipal Progress. Rev. ed. New York: The Macmillan Company, 1918.

INDEX

For Product Safety Concerns and Information please contact our EU representative GPSR@taylorandfrancis.com
Taylor & Francis Verlag GmbH, Kaufingerstraße 24, 80331 München, Germany

www.ingramcontent.com/pod-product-compliance
Lightning Source LLC
LaVergne TN
LVHW011120300926
849160LV00008B/457
* 9 7 8 1 1 3 8 4 9 1 3 6 6 *